D0216632

Project Planning, and Control

Also available from Taylor & Francis

Construction Project Management

Peter Fewings and Martyn Jones

Taylor & Francis Hb: 0–415–35905–8
Pb: 0–415–35906–6

Construction Safety Management Systems

Steve Rowlinson

Spon Press Hb: 0–415–30063–0

Human Resource Management in Construction Projects

Martin Loosemore, Andrew Dainty and Helen Lingard

Spon Press Hb: 0–415–26163–5
Pb: 0–415–26164–3

Occupational Health and Safety in Construction Project Management

Helen Lingard and Steve Rowlinson

Spon Press Hb: 0–419–26210–5

The Management of Construction Safety and Health

R. J. Coble, T. C. Haupt and J. Hinze

Spon Press 90–5809–328–X

Information and ordering details

For price availability and ordering visit our website **www.tandf.co.uk/builtenvironment.com**

Alternatively our books are available from all good bookshops.

Project Planning, and Control

David G. Carmichael

LONDON AND NEW YORK

First published 2006
by Taylor & Francis
2 Park Square, Milton Park, Abingdon, Oxon OX14 4RN

Simultaneously published in the USA and Canada
by Taylor & Francis
270 Madison Ave, New York, NY 10016, USA

Taylor & Francis is an imprint of the Taylor & Francis Group

Typeset in Sabon by
Integra Software Services Pvt. Ltd, Pondicherry, India
Printed and bound in Great Britain by
MPG Books Ltd, Bodmin

British Library Cataloguing in Publication Data
A catalogue record for this book is available
from the British Library

Library of Congress Cataloging in Publication Data
A catalog record for this book has been applied for

ISBN 13: 9–78–0–415–34726–6
ISBN 10: 0–415–34726–2

Contents

About the author

David is a graduate of The University of Sydney (BE, MEngSc) and The University of Canterbury (PhD) and is a Fellow of The Institution of Engineers, Australia, a Member of the American Society of Civil Engineers, formerly a Graded Arbitrator with The Institute of Arbitrators, Australia, and a trained mediator. He is former Head of the Department of Engineering Construction and Management and currently a Consulting Engineer and Professor of Civil Engineering at The University of New South Wales.

He has acted as a consultant, teacher and researcher in a wide range of engineering and management fields, with current strong interests in all phases of project management, construction management and dispute resolution. Major consultancies have included the structural design and analysis of civil and building structures; the planning and programming of engineering projects; the administration and control/replanning of civil engineering projects and contracts; and various construction and building-related work. In addition there have been numerous smaller consultancies in the structural, construction and building fields. He has provided expert reports and expert witness in cases involving structural failures, construction accidents and safety, and contractual and liability matters.

He is the author and editor of seventeen books and over sixty-five papers in structural and construction engineering and construction and project management.

Preface

It is interesting to watch the development of a discipline over the years. There have been many contributors to project planning. The author was fortunate enough to learn the fundamentals from James Antill (see for example, J. M. Antill and R. W. Woodhead, *Critical Path Methods in Construction Practice*, John Wiley and Sons, New York, 1st ed. 1965). When *Construction Engineering Networks* (D. G. Carmichael, Ellis Horwood/John Wiley and Sons, Chichester, 1989) was written, it seemed a losing battle to convince practitioners to use network methods for planning. Some practitioners were convinced, but the majority of planners saw network methods in a negative light. And that was at a time when reasonably user-friendly computer packages were available. Today such packages are universally used, if only as a way of generating bar charts. The challenge now is to understand what planning actually is, and this is the basis of this book; people are going through the motions of planning but do not understand the fundamentals of their trade. As such, the book treads new ground.

A systems approach is adopted, with control systems terminology preferred for its precision and usefulness over conventional planning terminology. In particular, state space and related terminology of modern control theory is preferred.

The book assumes that the reader is familiar with elementary planning, as is undertaken on all projects on a daily basis. If used as an introductory textbook, supplementary material such as given in Carmichael (1989) would be useful.

A main thrust of the book is seen as sending planning thought in the correct direction.

Project management is now commonplace in most walks of life ranging from the business world, service industries to technological endeavours. Central to project management is planning. Most matters in managing projects, whether it relates to time, costs, resources and so on, involve planning. Planning is a crucial issue in project management and one on which many other aspects develop.

Existing treatments on planning jump head first into network analysis, both deterministic (CPM – critical path method) and probabilistic (for example, PERT – program evaluation and review technique, and Monte

Carlo simulation), with all refinements including overlapping relationships, and the use of industry-preferred packaged software. The author, amongst many, is guilty of such an uninspired and pedestrian approach (Carmichael, 1989), justified falsely by the reason of 'practicality'.

A more inspired approach is possible. In fact, planning can be shown to be a systems synthesis or inverse problem. As such, the solutions to planning problems are not unique. In most cases, planners are only after a satisfactory solution, or a solution that they can live with, and do not spend the additional time searching for the optimal solution. A planner may also be under time pressures to come up with quick solutions.

However, planners expediently reverse the logic and deflect attention from their inability to come up with best solutions, on-time pressures and pseudo 'practicality' arguments, when in fact planners do not understand the synthetic nature of their job. Planners are unaware of and do not understand the components of the synthesis problem, and so they never know where they are relative to the optimum.

The book presents a totally new approach to project planning, and relates this to existing treatments. The aims of the book are to understand project planning and to contribute to thinking on project planning. One of the overriding reasons for writing this book is to counter the myriad of misconceptions and thinking errors that exist among writers on planning. The level of thinking that goes into planning in many cases is very superficial and cookbook in nature. To counter this, the book adopts a systems view to provide a rigorous framework. Rigour in the usage of terminology is also stressed.

The book aims to develop the reader's professional skills and thinking in the planning component of project work, to enable the reader to perform more effectively, to understand project planning procedures and to gain an insight into the associated skills. Numerous case examples from diverse industries, and exercises support the approach.

The book is aimed at lecturers and students (postgraduate and undergraduate) in project management, and practitioners involved in projects (all industries).

It contains case examples and exercises and hence could be used as a textbook.

Organisations, for which the material is applicable, include universities, professional organisations in engineering (all types), architecture, building and project management.

Nothing equivalent exists. Currently most people use project management texts that have planning (and 'control') as one component. The book rethinks planning practices, and outdates thinking in existing texts.

Acknowledgment

The book contains a number of case examples contributed by many people. Their contribution is gratefully acknowledged.

Notation

Continuous time version

t	time
t_S	project start time
t_F	project finish time
t_{now}	present time
$u(t)$	control at time t
$x(t)$	state at time t
$z(t)$	output at time t
$\zeta(t)$	disturbance at time t

Discrete time/staged version

k	period/stage counter, $k = 0, 1, 2, \ldots, N{-}1$
N	number of periods/stages
$x(k)$	planned-for state after period/stage $k{-}1$
$x(0)$	initial state
$x(N)$	final or terminal state
$xA(k)$	actual state after period/stage $k{-}1$
$x'(k)$	revised planned-for state after period/stage $k{-}1$
$var(k)$	variance; difference between planned-for state and actual state after period/stage $k{-}1$
$u(k)$	control at period/stage k
$\zeta(k)$	disturbance; anything that prevents $x(k+1)$ being obtained exactly

Activity information

TF	Total float
FF	Free float
IF	Interfering float
EST	Earliest start time
EFT	Earliest finish time

LST	Latest start time
LFT	Latest finish time
LT	Lead time period
SOS	Sum of squares

Part I

Conventional treatment from a systems perspective

Chapter 1

Introduction

Background

Many people are aware of the humorous paradoxical sign:

It also reflects a common saying, but can you plan any way other than ahead? In a like vein, people also refer in a redundant fashion to 'future plans', or if a project is not going well, it is said to 'lag behind' schedule.

Popular adages of planning

There are a number of adages and popular quotations involving planning:

In all matters, success depends on preparation. Without preparation, there will always be failure. (Confucius)

Action without planning is the cause of every failure.

Mediocrity is so easy to achieve, there is no point in planning for it.

If you fail to plan, you plan to fail.

If you don't know where you are going, you will end up somewhere else. (L. Peter)

If you don't know where you are going, any road will take you there.

The nicest thing about not planning *is that failure comes as a complete surprise, rather than being preceded by a period of worry and depression.*

4Ps Poor planning = poor performance.

6Ps Prior preparation and planning prevents poor performance.

6Ps Prior preparation prevents poor performance – probably.

Even if you do nothing, it should be done on time.

Plan the work, and work the plan.

If you plan to fail and you do, then you have been successful.

Real life is what happens when you make other plans.

The stability of a plan tends to vary inversely with its extension.

Planning also is mentioned in children's fables. For example, one of Aesop's fables gives the story of the ant and the grasshopper, where the ant is industrious collecting food over summer while the grasshopper plays, only for winter to come and the grasshopper goes hungry, leading to the conclusion that 'it is best to prepare today for the needs of tomorrow.' Aesop's fable of the cat and the fox, where the cat knows one trick to escape the dogs and uses it, while the fox knows many tricks but cannot decide which to use and gets caught by the dogs, concludes that 'one good plan that works is better than a many doubtful ones.'

Two noticeable things emerge from these adages and fables, and other publications on planning:

- Everyone says that there is a need for planning, and a need for genuine effort to be put into planning.
- Everybody has a different idea of what planning is.

Therein lies the source of most of the troubles preventing the advancement of the understanding of planning. To discuss these in a project context, it is first necessary to take a few steps backwards, and clarify some thinking. The first step relates to the distinction between a project and its end-product.

Project versus end-product

A project will commonly come about because of an identified need or want for some product, facility, asset, service and so on. This end-product is achieved through a project (Figure 1.1).

This distinction sometimes causes people confusion and many people are not aware of the distinction, or of the need to make a distinction. For example, people sometimes refer to a building as a project. It is not. The processes that go together to materialise the building are the project. The building is the end-product. (It is acknowledged that the definition of a project is sufficiently flexible to include the operation and maintenance phase of a product within what is called the project. However, this is not the issue here.)

There will be objectives and constraints relating to the end-product, and objectives and constraints relating to the project. (The plural terms objectives and constraints are generally used in this book even though the singular forms may be applicable in some situations; the exception is where the singular is deliberately intended.) The former influences what the end-product will look like and how it functions. The latter influences the way of getting to the end-product. The term 'objective' here is used in a correct systems synthesis sense, not as used loosely by lay people (Carmichael, 2004).

The objectives and constraints are derived from the value system of the project/end-product owner. They lead to establishing the scope, and planning practices.

Later chapters develop a hierarchy of objectives and constraints – at project, activity, element and constituent levels. A project is composed of activities. An activity is composed of elements. Elements are composed of constituents (Figure 1.2).

Planning and design

Following the distinction between an end-product and the means to the end-product, design and planning functions are seen as parallel and fundamental undertakings on projects. Broadly on a project, Figure 1.3 applies.

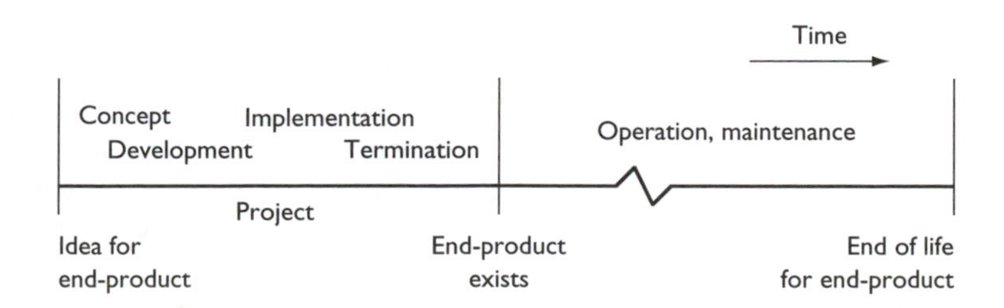

Figure 1.1 Project and end-product timeline.

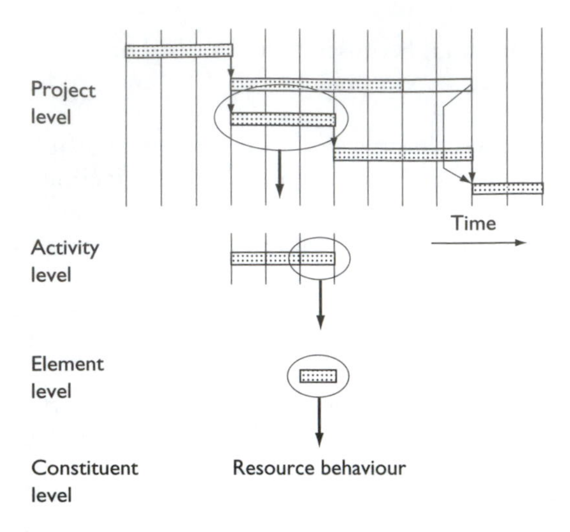

Figure 1.2 Hierarchical levels in a project.

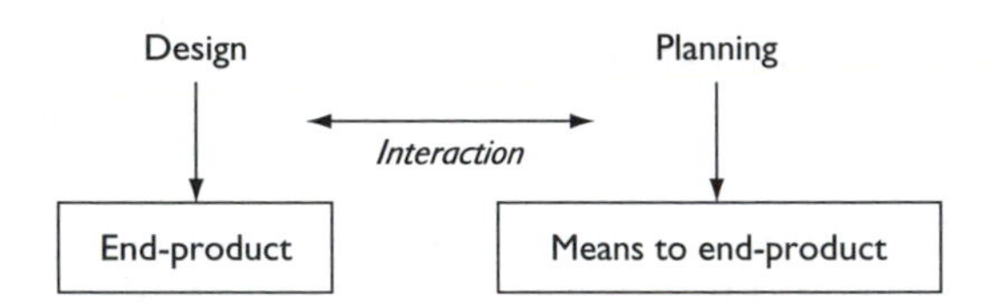

Figure 1.3 Fundamental project undertakings.

Both the planning and design problems are shown later to be inverse (synthesis) systems problems, although popular practice is to solve these in an iterative analysis fashion.

It is interesting to note that in nearly every project management text, there is a section on planning, and invariably the text authors feel obliged to convince the reader that there is a need for planning and of the benefits of planning, and to justify the pages allocated to planning. The same holds for courses given on planning. This indicates the immaturity of project management as a discipline and as practised. By comparison, design texts and design courses do not have to convince readers and practitioners of the need for design; it is accepted as fundamental in order to get to the end-product. Why then is not the parallel project task of planning accepted as fundamental, and let the discussion get straight on to the best way to do it?

Also of interest to note is that no owner would let a person, who has no design education, design something for them. Owners are very discerning when it comes to design. Yet anyone can get away with calling him/herself a planner without any planning credentials, and undertake planning for that same 'discerning' owner. Yet design and planning are both fundamental project tasks on which the project outcome equally depends. It would be nice to think the reason for this was that the owner has to live with the visual impact and functionality of the end-product, but can soon forget about the cost and duration impact of getting to the end-product. But the real reason appears to be the different maturity levels of design and planning, or more appropriately the immaturity level of planning compared to the maturity of design. It may also be that planning is an order of magnitude more difficult than design, yet it is practised by people with specialist knowledge in the inverse ratio. People have incompatible standards and expectations when it comes to design and planning work.

Another form of undesirable imbalance can be observed in some project managers and in some project management publications which take the view that activity scheduling is project management; if you have a computer package that does scheduling, then you have all that is needed to manage a project.

The false belief in the need for planning

Everyone says that there is a need for planning, and a need for genuine effort to be put into planning; the latter is acceptable (it is not unreasonable to expect that effort put into planning improves the chance of project success) but the former implies that no one recognises planning as a fundamental project task. Since planning is a fundamental project task, talk of need for planning is off the mark, as is the question 'why plan?' posed in books. Planning is carried out (to varying degrees of sophistication) on every project whether people realise it or not. Without planning, a project doesn't proceed. Perhaps when people talk of a need for planning, they are talking about structured planning, as opposed to *ad hoc* planning, rather than talking of the no-planning versus some-planning scenario. But it is more likely that people don't understand the place of planning.

Whether people are involved with projects or not, everyone carries out planning in some form covering the tasks that go to make up their daily lives. This applies to their home roles, social pastimes, commuting between home and work, and work tasks. Having breakfast and getting dressed in the morning, driving from A to B, spending a weekly pay cheque etc. are all planned to some degree. The planning might be casual or very detailed; it might be well thought out or poorly thought out; it might be over short or long time horizons. Without planning, the task or project does not proceed. Different planning leads to the task or project being carried out differently.

What is planning?

Everybody has a different idea of what planning is. See the above adages, for example. Project management texts give a variety of definitions. And if lay persons, practitioners and texts on 'strategic planning' and 'market planning' are consulted, it appears that the word planning is used by most people in the sense of Alice in Wonderland:

> 'When I use a word,' Humpty Dumpty said in a rather scornful tone, 'it means just what I choose it to mean, neither more nor less'.
>
> 'The question is', said Alice, 'whether you can make words mean so many different things'.
>
> (Lewis Carroll, 'Through the Looking Glass', Ch. 6)

Some people refer to planning as 'time management', and vice versa, but this is too restrictive, and not correct. Strictly, time cannot be managed.

In broad terms:

Planning concerns itself with the 'means to end-product' problem. Planning establishes how and what work will be carried out, in what order and when and with what resources (type, and number or quantity) (additionally expressed in a money unit).

The parallel matter of *design* concerns itself with the 'end-product' problem.

(This last sentence refers to the design function. Of course, the actual activities involved in the design process can be treated as a subproject in its own right and planning applied to these activities. Similarly, for those who like embedded problems, the activities involved in the planning process can be treated as another subproject in its own right and planning applied to these activities. This sort of logic can be continued ad infinitum, for those so inclined.)

'How' implies 'method'. 'Order' implies 'sequence'. 'Resources' are people and equipment. Resource type, number (quantity) and production are associated with method, and order and timing (start, duration and finish) of activities. Resources (type, and number or quantity) are typically expressed in a common measurement unit of money (Figure 1.4).

Planning outcomes provide information, for example, on:

- Schedule matters (project completion date, milestones, differentiation between important (critical, Pareto, . . .) and not important activities, . . .)
- Resource matters (type, number or quantity, and timing requirements, including outsourcing)
- Production matters (related to schedule)
- Money matters (project cost, cash flow, . . .).

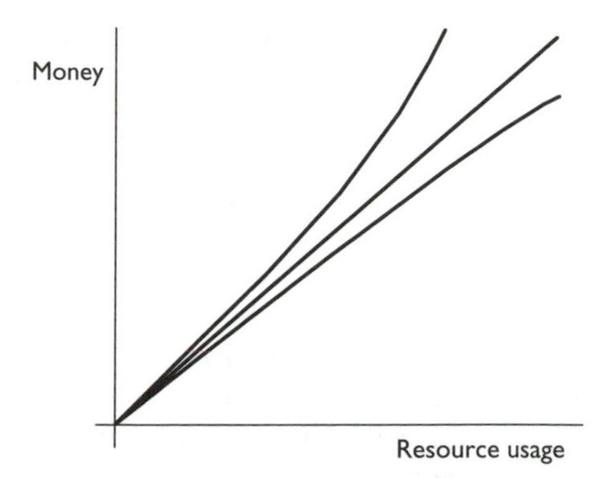

Figure 1.4 Some possible relationships between resource usage and money.

All tend to be strongly related. Oscar Wilde observed in Phrases and Philosophies for the Use of the Young that, *Time is waste of money.*

Prophecy

Planning is distinguished here from prophecy.

> It is essential for thinking man to believe in free will, to believe that his future can be changed by thought and action. Prophecy denies this and declares that all futures are immutable and fixed, that nothing man strives towards is relevant, as is already preordained by whatever governs his future, be it God or Destiny.
>
> ... But on the other hand, the answer may lie in Dunne's concept of alternative variable futures which run parallel to each other. Each man receives the future dependent upon his earlier actions.
>
> (Extracted from E. Cheetham, The Prophecies of Nostradamus, Corgi, London, 1975, pp. 14–15)

The planning function

In systems terms, planning establishes the value of *control* (input, action, decision), $u(t)$, throughout the project duration, $t_S \leq t \leq t_F$. $u(t)$ has multiple parts (that is, it is a vector quantity of control variables) that contain information on method (including sequence), resources (type, and number or quantity and hence money) and resource production rates (or equivalent). t is time. The project is viewed as a dynamic system.

$$u(t) = \{\text{method}(t); \text{resources}(t); \text{resource production rates}(t)\}$$

A hierarchical and staged approach to planning might be adopted. (Four main hierarchical levels are considered in this book – project, activity, element and constituent – Figure 1.2.) Decoupling the hierarchical levels and stages facilitates the solution, but introduces iterations. Determinism would generally also be preferred by planners.

In general, there are many possible solutions (values of control variables) to the planning problem. The associated objectives enable a choice between alternative solutions to give that solution which is optimal or efficient in resource usage, most economical or best in some other way, for example minimum duration. That is, the control is chosen so as to extremise (optimise/minimise/maximise) the objectives while satisfying any constraints present. Economy here implies an objective involving money; money (in total or as a function of time) may also be present as a constraint.

Objectives and constraints, where mentioned below, relate to the project unless otherwise noted; the end-product problem is not considered in depth in this book. Carmichael (1981) considers the end-product problem.

Planning controls are selected over the full time frame of a project for a globally optimum solution. Planning separately on a stage-by-stage or even day-by-day basis is suboptimal. Control selections early in a project constrain future updated control selections.

$u(t)$ is selected based on whatever objectives and constraints are present. Frequently a dominating constraint is that of a desired project performance or behaviour (later referred to as state $x(t)$) over the project duration, $t_S \leq t \leq t_F$; such a constraint may eliminate most of the potential for optimisation. The selection implies knowledge of work practices, and (fore)thought as to what might happen later in time. In this last sense, planning is said to be proactive; the alternative is to be reactive and not carry out forethought as to what might happen. $x(t)$ has multiple parts (that is, it is a vector quantity of state variables) that contain information on cumulative resource usage, cumulative money usage or cumulative production or equivalent. (Project output, $z(t)$, is also an indicator of project performance, but for the stripped-back project formulations given later, output and state are the same.)

Flow-on aspects to assist management

This book pushes a rational and structured approach to planning as being preferential to an *ad hoc* approach.

Planning well-done gives rise to many flow-on aspects that assist the overall project management. For example:

- Critical/important work items can be highlighted. Attention can be focused on those activities that are critical in terms of achieving a desired project completion date, or Pareto in some other sense.

- All activities necessary for the project are delineated.
- Potential trouble areas and obstacles are identified.
- Communication between project personnel can be assisted. Project team members and other project participants can be informed of their duties and when they are to contribute to the project and for how long.
- Planning outcomes can provide a basis for coordinating the work of project participants.
- Coordination between simultaneous projects can be facilitated.
- A more satisfying work environment for project personnel can be obtained where it can be seen that proper planning has taken place. Stakeholders can be impressed.
- Uncertainties affecting the project can be reduced. Allowances and contingencies can be reduced. Risk management practices are implicitly incorporated, without needing to mention the term 'risk management', or to separately go through the risk management process. Predicting the influence of changes and delays is enhanced.

Flow-on aspects to please the project owner

Project owners prefer planning well done because of the improved possibility of a project finishing by a desired date and within a given budget. Owners can be provided with information regarding dates and costs (throughout a project, together with that at project completion), information necessary to conduct their total affairs. The impact of changes, delays and uncertainties associated with completion dates can be ascertained.

Case example – Desktop computer software

One organisation, with branches distributed over a large geographic region and in a number of countries, decided to standardise on an operating system and applications used by staff on their desktop computers. A consultant was contracted to provide the redesign and specification. One particular personal computer brand was selected as the hardware.

The branch offices were geographically dispersed and the type of problems differed from country to country, making planning difficult:

- Additional funds had to be allocated, taking into account technical uncertainties and equipment availability; for example a computer may cost twice as much in one country as another.
- In some countries, additional funds were required to ensure certain officials approved of the private network infrastructure.

- While performing the installation of the equipment, there were intermittent power failures that resulted in the installation having to go back to the start of the procedure.
- Some local vendors caused some logistics problems that delayed the project schedule. A vendor would promise equipment to be delivered based on a contract, but then not do so.
- Delays were common in some countries.
- Certain specified hardware and software was unavailable in some countries. Equipment had to be sourced elsewhere. An additional time period and cost was involved.
- Even if the equipment arrived on schedule, unforeseen problems occurred during the installation of the servers.
- No provision was made for additional parts/spares for replacement. This caused delays, which meant that other sites had to be rescheduled as well.
- There was a lack of skilled technical personnel. This required staff to be retrained in order to do the implementation and provide support after implementation.
- Additional staff had to be sourced from outside the organisation to assist in the implementation.

In hindsight, an additional time period and cost should have been included in the planning for this project. With all the problems encountered, an additional year and a half was required for the project over that originally planned – getting the infrastructure installed, equipment delivered and staff trained.

Replanning

Generally the term 'control' is used loosely in the literature and by lay persons and practitioners. Particularly at fault is the usage in some computer packages, where package marketing promotes the feature of 'control', but really refers to something else, in many cases something no more sophisticated than handling data. In other cases the term is used to mean placing some limit or containment on something such as costs, or keeping this something from growing unacceptably. In general management texts, and lay terms, the term 'control' is used in the context of power, dominance or influence. (Something that is 'out of control' is unable to be influenced.) It appears that the word 'control', like planning, is used by most people in the sense of Alice in Wonderland:

> 'When I use a word,' Humpty Dumpty said in a rather scornful tone, 'it means just what I choose it to mean, neither more nor less'.

> 'The question is', said Alice, 'whether you can make words mean so many different things'.
>
> (Lewis Carroll, 'Through the Looking Glass', Ch. 6)

In this book, the term *control* is used in a systems sense, meaning *the actions, input or decisions of the planner (decision maker) in order to bring about a desired project (system) performance.*

The popular usage of control as stopping something from growing unacceptably or containing something is incorporated through the use of constraints in the synthesis formulation.

> The common terms 'project control', 'project control system', and 'cost control system' are not favoured in this book because their lay meanings have a tendency to mislead; their sloppy usage is inconsistent with the optimum control systems usage preferred in this book. Rather, the term *replanning* is preferred to 'project control'.

What replanning, when necessary, does is come up with revised values of control variables, u(t), for the remainder of the project, $t_{now} \leq t \leq t_F$. The nature of u(t) is unchanged – multiple parts (vector quantity) that contain information on method, resources and resource production rates.

Replanning does what is essentially embodied in Bellman's principle of optimality, which was devised for different purposes. The principle states: '*An optimal policy has the property that whatever the initial state and initial decision, the remaining decisions must constitute an optimal policy with regard to the state resulting from the first decision.*' The words 'decision' and 'control' may be interchanged. See Carmichael (1981, 2004).

In essence, replanning is no different to planning. It is planning carried out subsequent to any original planning, using updated initial conditions, shortened time horizon, updated constraints and possibly updated objectives. As such, most of the discussion on planning applies to replanning.

Project performance (the project state x(t)), is monitored on an ongoing (usually regular) basis, *compared* with planned performance (typically in a report) and, if necessary, *corrective action* is taken to send the project in a desired direction (on a desired-state trajectory). The controls bringing about corrections may be in the form of method (including sequence), resources (or money), and/or resource production rates (or equivalent) input to the project. The controls are chosen to extremise the objectives while satisfying any constraints present. *Applying changed or different controls amounts to replanning*. Accordingly, planning (in the form of replanning) might be referred to as an ongoing or evolutionary process. The 'planning phase' of a project occupies the total project duration.

Generally, it can be said that prior work done constrains future (updated) planning actions. Hence the constraints are changing as a project progresses. The flexibility in choice of controls decreases with time.

Monitoring and reporting are necessary precursors to knowing whether replanning is necessary or not. A changing project and its environment (represented later by disturbance, $\zeta(t)$) lead to the need for continual monitoring and reporting of project progress. Work rarely is done exactly as planned. (Project) scope also rarely stays constant from project start to project end. Objectives and constraints may also change.

Planning is based on certain assumptions, and at project implementation, conditions might be different. As well, it is rare that any project as originally planned is brought to fruition. For example:

- There are inevitable changes relating to technical specifications, method of implementation and owner needs (revised priorities, rework, . . .); the scope (of work) changes; variations occur.
- The original planning may have been unrealistic, unachievable, inaccurate or contained mistakes particularly with regard to activity duration estimates, activity relationships and activity resource demands and production; activity sequencing used for planning may have been incorrect; duration and/or cost estimates may have been too optimistic/pessimistic.
- There may be vendors/suppliers or contractors who are unable to meet their original target dates, or perform to expectations/requirements; subcontractors may be unknown.
- Extreme unexpected events may occur including unimagined geotechnical conditions; there may be delays due to industrial trouble and bad weather.
- There may be changes in design; the design may not have been checked thoroughly.
- There may be changes in the availability of resources (people and equipment); equipment may turn out to be not suited to the task; sufficient resource numbers are unavailable when required.
- There may be changes in the availability of money.
- Material prices may change; there may be changes in the availability of materials.
- There may be technology changes or technical difficulties; the use of new or previously unused technology may not have been thought through fully; quality or reliability issues occur.
- There may be coordination problems with other groups; necessary preceding activities may be incomplete.

- There may be access, space or logistics problems.
- The market, the economy, competitors, regulations and interest rates may change.

And according to Murphy's Law: *Anything that can go wrong, will go wrong*.

Case example – Bulk concentrated food mix

The project involved developing a new plant to manufacture a bulk concentrated food mix for industrial use. The capital component of the project included civil work comprising of a new building to locate a bulk sugar storage and handling plant and an addition of a mezzanine floor to house machinery in an existing building. The mechanical work comprised the installation of product silos and associated valving and pipework, installing new processing and packaging machinery and installing a new automatic cleaning system. The electrical work comprised the electrical installation for the above work and implementing software for the automatic control of the mechanical plant via programmable controllers and computer screens. The total budgeted cost of the project was several million dollars but overran by a third. The time span for the project was estimated to be about a year, but also overran by a third. The civil work was managed by a professional project management company and executed by contractors. The mechanical and electrical design and installation was managed by in-house project engineering staff.

This project had both overruns in cost and duration. There were uncertainties in the planning that resulted in delays in the implementation stage. These are described below.

In the initial planning, there was no clear estimate of the required capacity for the new plant. When the project was initially budgeted, the return on investment was calculated on sales figures based on potential in the market, that is subjective estimates. The plant was sized based on the forecast budgeted sales figures plus a growth factor. However, later in the project, the plant was required to be of a significantly larger capacity. This affected the plant layout, and the detailed design could not begin until new dimensions, power requirements and prices for machinery were obtained. The delivery periods were also longer for plant and equipment.

There was a resulting problem in the location of the sugar store and handling equipment. The architectural design for the building and the layout had been done based on the original estimates, but there was inadequate space for the new plant size. This resulted in further delays in redesigning the building and obtaining development approval from the local authorities.

There was also no certainty in the production process because the production technique had to be changed for commercial and contractual reasons. The plant was basically of an unproven or novel design that was scaled up from the research and development (R&D) laboratory trials. During the commissioning stage, a number of modifications had to be done to achieve the required product specifications. These modifications were done on a trial-and-error basis and consequently added to the increase in the duration and cost of the project.

There was no project manager for the entire job. The person responsible for the project was the project engineering manager, but he was supervising a number of other projects at the same time. There was a project management company that looked after the civil construction, while three mechanical engineers and one electrical engineer undertook the design and the supervision of the plant installation and commissioning. Again, these engineers were associated with other projects at the same time. Some sections of the project were completed well in advance, while others had significant delays mainly due to long delivery time periods of some plant and modifications to the design.

Case example – Land development project

In a land development project, activities included work in dredging (tidal river), dewatering bunds, dry bulk earthworks, bridge piling and piers, rock foreshore protection, sewerage and stormwater drainage. The original planned construction duration was about a year.

In the initial planning, uncertainties recognised included those associated with: weather and flooding, encountering marine clay layers during dredging and dry excavation, time periods for sequential approvals by government departments of various components of construction, and contractor performance relevant to schedule and cost.

Undefined material in dredging The chance of encountering marine clay layers, unable to be dredged, created an uncertainty in the project duration and cost. In the initial planning, the presence of mud was not recognised. Available boreholes represented samples some 200 m apart, and previous experience from construction in this particular area had indicated that no dry marine mud would be encountered and was not allowed for. As the project progressed, a significant quantity of mud was found in one area. The response to this was to adopt an alternative method of material removal. Material was trucked, in lieu of dredging, at an additional cost and an additional duration.

Approval time periods The uncertainty of the approval time period of subactivities by local authorities was recognised in the initial planning. However, the duration of the delays in achieving approvals was grossly underestimated. Changes in environmental legislation had placed more stringent requirements (for example, additional groundwater testing) since commencing the project. For this particular subarea, access was not available until approval had been issued. This influence was handled by arranging the contractor to carry out advanced work in other approved areas for future stages. In this manner its resources were not left idle (a potential cost to the project without production). Although achieving meaningful work in the future stages, the owner's cash flow troubles caused considerable internal management difficulties.

Contractor performance The production performance of the contractor was well known for this type of work, and the estimated activity durations were reflective of expected performance. In the construction phase, there were no difficulties in this respect. While the tendered project duration appeared reasonable to the consultant and the owner, there still remained uncertainty in the activity duration estimates. This was particularly relevant to delays from a flood event.

Information from monitoring, a comparison of actual project performance with planned project performance, as well as absolute project performance, is processed into *progress reports*, which are used as the background to replanning (Figure 1.5).

One useful report is that which shows together all *variances*, that is the difference between as-planned performance and actual performance for all indicators of project performance (that is, state). A generic form of collective variance reporting is shown in Figure 1.6. Figure 1.6 shows a two-dimensional *state space* (a space with state variables as axes), but the same holds true for where the performance of the project is reflected in more than two states; the space becomes a hyperspace for dimensions greater than three. (For dimensions greater than three, tabular reporting of these variances, or pairwise reporting of these variances might be preferred.) Two state trajectories are shown in Figure 1.6 – one corresponding to the as-planned behaviour, and the other corresponding to the actual behaviour. The variances in the states are also shown.

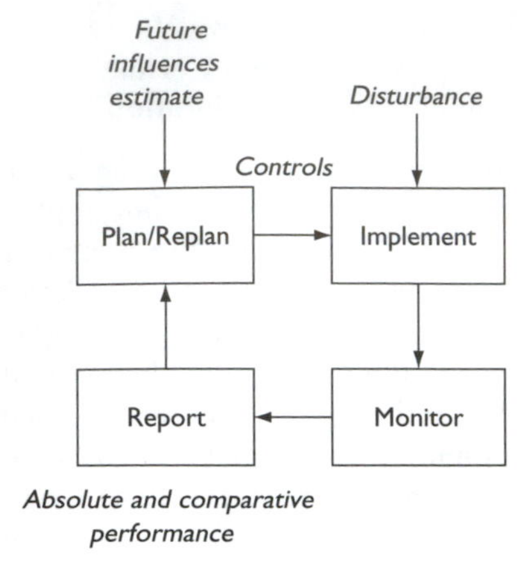

Figure 1.5 The replanning process.

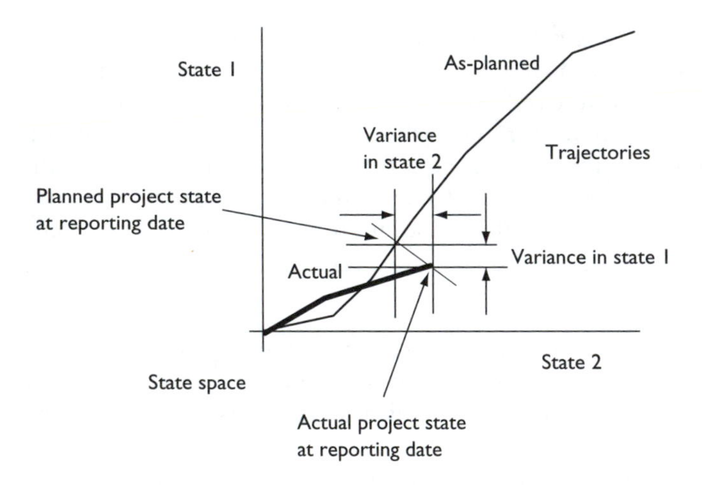

Figure 1.6 Generic way of representing project variances in a state space.

As an example, the two states in Figure 1.6 might be cumulative resource usage and cumulative production. The variance in cumulative production will indicate whether a project is 'ahead/on/behind schedule'; production performance not only reflects actual resource production rates, but may also reflect delays, and popular reporting may be more accustomed to schedule and delay reporting rather than reporting as in Figure 1.6. The variance in cumulative resource usage, when converted to a money unit, will indicate whether a project is 'over/under budget'. Popular reporting may also be more accustomed to cost reporting rather than reporting as in Figure 1.6; in that sense it might be preferred to aggregate the cumulative resource usage states across all resource types and express this as a cumulative cost state. The variances from Figure 1.6 might then be summarised over time in diagrams such as Figure 1.7 and 1.8 to give a historical record of project performance.

A way of reporting variances, less useful than Figure 1.6, is shown in Figure 1.9, where individual states are plotted against time. This only gives a part picture of what has happened on the project up to the reporting date. For example, where cumulative resource usage is the state being represented in Figure 1.9, the figure may indicate that less resource numbers have been used than had been planned, but says nothing as to whether production is as-planned, ahead of as-planned or behind as-planned.

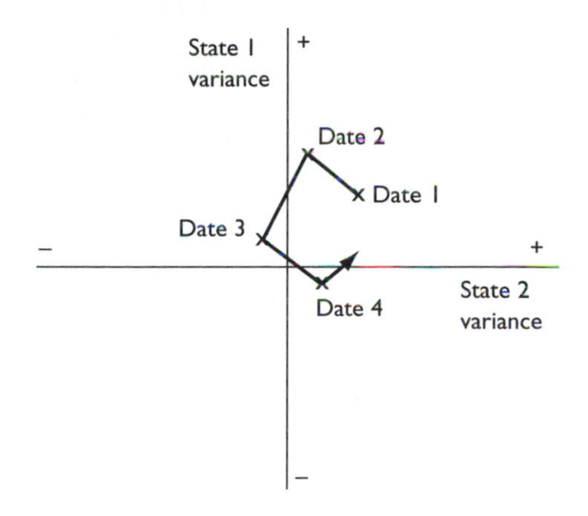

Figure 1.7 Progressive plot of state variances.

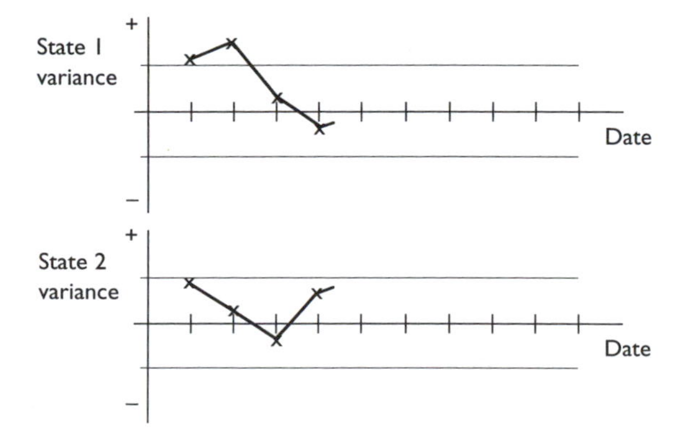

Figure 1.8 Separate plots of state variances.

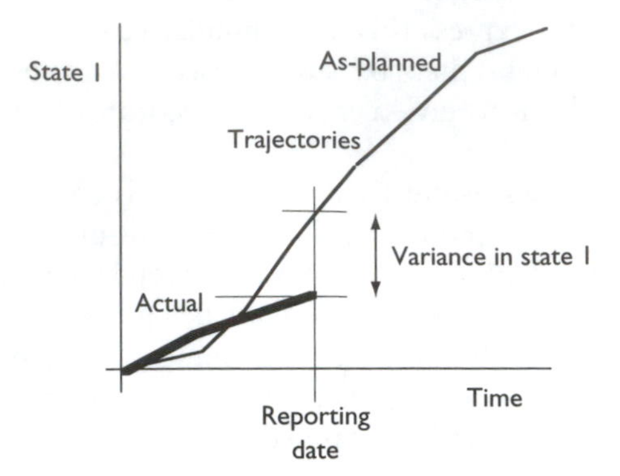

Figure 1.9 A way of showing project variances less useful than Figure 1.6.

Earned value thinking introduces a modification to Figure 1.9 to make it more useful (when money and production/schedule are being looked at), but still Figure 1.6 is a preferred form of representation.

Example replanning controls

The replanning problem, in principle, is no different to the planning problem. The available controls are the same in both cases; replanning may differ in that it becomes an adjustment or change to previously selected controls from the earlier planning. Possible replanning controls are many, ranging from altered manning numbers through bonus systems, preventative maintenance etc., but commonly centre around method, the use of resources (implying money), and resource production rates (or equivalent), for example:

Project level controls

- Resources may be redeployed, redistributed, increased or decreased within the project. This could be expected to alter the critical paths and the scheduling of activities and costing.
- Using new/alternative methods (changed work method), and resources for the remainder of the project. This implies a new network and new costing.
- Fast-tracking the remainder of the project; activities done in parallel rather than in series; overlapping relationships between activities.
- Expediting materials; introducing 'just-in-time' or similar inventory practices.
- Outsourcing part of the work.

Activity level controls

- Compression (shortening durations) of activities leading to network compression
- Splitting activities
- Trading off resource numbers and duration
- Changing the resource work calendars.

Element level controls

- Working extra shifts
- Using extra resource numbers.

Constituent level controls

- Providing incentives
- Altering resource production rates.

This list is not totally exhaustive. Controls selected depend on the planner's experience and expertise, as well as the situation and some creativity. Some solutions may be better than others. The best solution is that which extremises the objectives while satisfying any constraints present. However, generally planners are after a satisfactory solution rather than some theoretically optimum solution.

Work study and value management (including constructability/buildability) studies can be useful here, though all these are subsumed by a systematic approach to problem solving (Carmichael, 2004).

Replanning may also involve the removal of constraints (physical, technical, logic, . . .) affecting the work.

Through all planning and replanning discussions above, it is assumed that estimates and assumptions are correct, which may not be the case. Correcting these will also affect the solution.

Planning effort

As with many matters involving effort and associated cost, there is a balance between overdoing and underdoing. Little planning saves money and effort on the planning work but may be wasteful in terms of the total project outcomes (extra cost, late completion, damages, . . .). Further effort and money put into planning may improve the project outcomes. Still further effort and money put into planning may, however, bring no additional benefits or savings in total project costs.

Unfortunately there is no reliable data on the cost (and amount) of planning. One reason for this is that people cannot agree on what actually constitutes planning and where the boundary between planning and other activities should be drawn. Sloppy usage and understanding of the term 'planning' by many people contributes to this. The cost (and amount) will also vary from industry to industry and organisation to organisation. Smaller projects may bear a proportionally higher planning cost than larger projects.

There is even less data on the relationship between planning effort and project outcomes. Carmichael (2004) argues:

> A project contains an immense number of influencing factors ([matters] that contribute to bringing about any given result), none of which is possible to be isolated in order to establish its influence on a project's outcome. On any given project some factors may be present, some may not be present, while some factors may be more dominant than others.
>
> It is not possible to demonstrate in any objective way how a particular factor contributes to project success. (. . . the inverse investigation problem cannot be solved.) There is only experiential observation or anecdotes of projects to support popular contentions. To that could be added 'gut feel', judgement and perhaps logic; the contributions of various factors seem reasonable propositions. Experience says that the contribution of particular factors reduces/increases the chance of a project under-performing.
>
> Objectivity is not possible because, although many projects are similar, no two projects are the same. A 'control' project cannot be set up, against which other projects with different factors are compared; the closest that may be obtained is to observe two similar projects but with different factors, but this is inconclusive. Each project cannot be done multiple times – each time with a changed factor. Medical experiments on people, it might be argued, suffer the same problem, but get around the issue to a certain extent by conducting trials on large numbers of people. Conducting trials on large numbers of projects, each containing multiple factors that have the ability to influence the projects' outcomes, is not feasible. Individual factors cannot be varied one at a time, while all other factors are kept constant. There is not a one-to-one relationship between any project inputs/factors and project outputs; to a certain extent, a project (as a system) is almost uncontrollable in systems engineering (Kalman **controllability**) terms; altering any one input/factor affects multiple outputs. There are also **observability** (in the sense of Kalman) troubles. The benefits purported to be due to one factor may be due solely or partly to some other project factors.
>
> Performing one project [task] affects other [tasks], and the outcome of other [tasks]. And on large projects there are large numbers of [tasks], most of which are interdependent.

An alternative is to approach the argument from the double negative perspective, namely by reasoning that with a certain factor, then project outcomes will almost certainly be undesirable. And this is supported by many case projects that have performed badly with such a factor. The consequent conclusion is drawn. Again this is reasonable, but can't be proven.

A number of studies have been reported trying to demonstrate a link between factors and project success. Typically they involve examining multiple real projects, and not any deliberate experimental projects. Projects are categorised, in a pseudo-rigorous fashion, as high, medium or low in terms of the factors and the (falsely drawn) conclusion then follows that there is a positive correlation between a factor and ultimate project success. The studies give an air of objectivity, but remain largely subjective.

It is difficult to establish what is the optimum (the balance between the amount of planning and project outcomes) planning effort. It is not even safe to observe projects and assume that practitioners have homed-in on the optimum by trial and error.

Case example – Food production

It was common practice within one food production company to undertake projects with much fervour and zeal. Often projects were fast-tracked for no apparent reason. Projects, which resulted from operational or competitive necessity, were also done at such a rapid rate that all conventional planning and project staging methodologies were ignored. The bottom line result of this was 'over budget, over schedule' projects, the bane of any project manager.

Engineering-based projects evolved from a production department, usually the production manager, with inappropriate costing, little associated planning and a timescale that could not be adhered to.

The shortfall in planning effort also reflected the insufficient knowledge or experience about the job and translated into a poorly constructed job specification that led to a shallow scope (of work) and ultimately unrealistic tender documents.

People within the accounting section at the company often argued that thorough planning was costly and could not be justified. It was felt that thorough planning also led to an increase in project duration and was only a way for some employees to ensure their employment.

This belief, in some ways, promoted the use of fast-tracking in order to circumvent thorough planning.

Project close-outs

Special planning commonly goes into the commissioning and handover part of the termination phase of a project.

Work in this phase ties together a lot of loose ends, prepares for the transition to take the facility to the operating and maintenance stages, analyses project performance, files all project data, disposes of or transfers resources including people and closes accounts.

Project personnel are aware of the need for thorough planning in the initial stages of a project but some neglect the planning of the project termination.

Some projects engage a new project manager and team specifically to cater for the commissioning and handover. This is then viewed as a 'project' in itself and planned accordingly.

Personnel problems, as the project winds down and people are transferred or laid off, are of particular importance during project close-outs.

Book outline

The book develops the subject in two parts:

- Part I – Conventional treatment from a systems perspective
- Part II – Synthesis treatment.

The reason for this is that the systems ideas presented in the book are new to planners. Part I takes planning practices (both good and bad) familiar to most planners, but couches them in the more rational framework offered by systems thinking. This elevates the practice of and thinking on planning. Part II offers new theoretical constructions for the planning problem, and although not intended to be used in everyday planning practice, offers considerable insight into what planning is. Planners will benefit by understanding true synthetic approaches. Selected topics in both Parts take a number of interesting planning topics and develop these in a fashion consistent with Parts I and II. Examples are included throughout the book to reinforce the subject matter.

Figure 1.10 shows a schematic outline of the flow of the book. Part II could be omitted in a first reading. Part I relies on the reader having some fundamental exposure to planning.

Terminology uses

Terms used interchangeably by writers include owner, client, principal, employer, developer, proprietor and purchaser. Generally, the term 'owner' is the one adopted in this book.

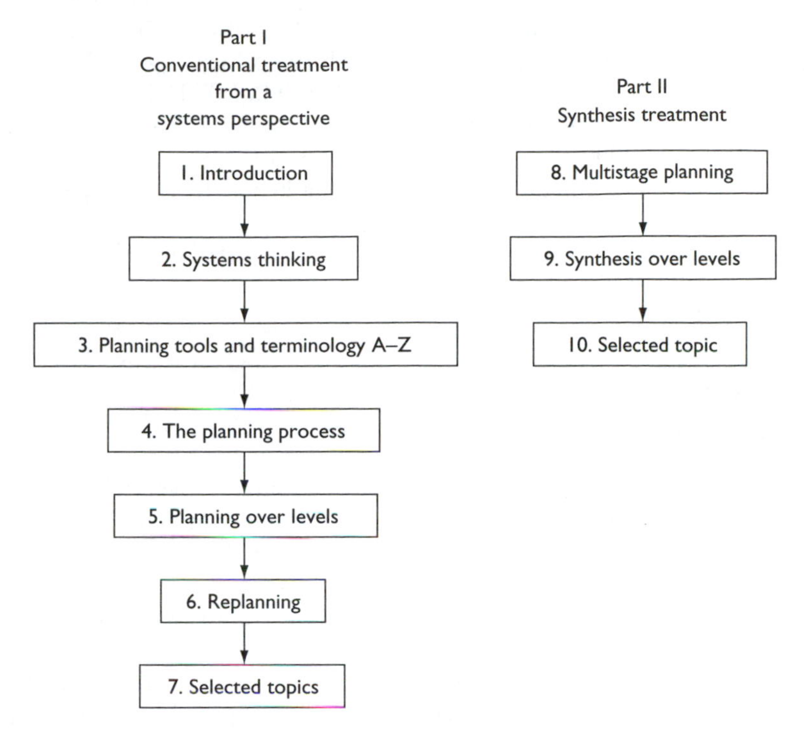

Figure 1.10 Outline of book.

Terminology used interchangeably in this book includes:

- State, output, performance, response and behaviour
- Control, action, decision and input.

Gender comment

Commonly references use the masculine 'he', 'him', 'his' or 'man' when referring to project personnel possibly because the majority of project personnel have historically been male. However, such references should be read as non-gender specific. Project management is not an exclusive male domain.

Influential publications

The following books by the author have ideas that were influential in the formation of the planning ideas advanced in this book:

Structural Modelling and Optimization, Ellis Horwood Ltd. (John Wiley and Sons), Chichester, 306pp., 1981, ISBN 0 85312 283 0.

Engineering Queues in Construction and Mining, Ellis Horwood Ltd. (John Wiley and Sons Ltd), Chichester, 378pp., 1987, ISBN 0 7458 0212 5.

Construction Engineering Networks, Ellis Horwood Ltd. (John Wiley and Sons Ltd), Chichester, 198pp., 1989, ISBN 0 7458 0706 2.

Contracts and International Project Management, A. A. Balkema, Rotterdam, 208pp., 2000, ISBN 90 5809 324 7 / 333 6.

Disputes and International Projects, A. A. Balkema, Rotterdam, 435pp., 2002, ISBN 90 5809 326 3.

Project Management Framework, A. A. Balkema, Rotterdam, 284pp., 2004, ISBN 90 5809 325 5.

Exercises

1. What is planning to you?

2. Look around you. The presence of planning is everywhere. Consider how people's routines are affected by the following end-products of planning:

- What would you do without television, radio and movie programs/guides?
- How would you like to live in a city without transport timetables?
- How could you study for a degree without subject, lecture and topic timetables?
- How would athletes achieve peak performance without a training schedule?

3. Consider a typical project. What form of planning might be undertaken by the various project participants – owners, consultants, contractors, subcontractors and suppliers?

4. Lecture timetables, transport timetables, television programs, training schedules, bar charts, time-scaled networks, and cumulative production plots are all examples of ways planning information is conveyed. Why do we use one way in one situation and another way in a different situation?

Examine some timetables, programs etc. and try to improve on the way the planning information is conveyed.

5. The question many people ask is 'why plan?' if plans are continuously changing. Also some people state that they have been associated with projects where no planning was carried out and the projects were successful. How do you respond to such comments?

6. List benefits of thorough planning. What are the disadvantages or drawbacks to thorough planning?

7. In what way can a plan limit flexibility in developing new ideas, stifle initiative, and focus people's attention on the small picture at the expense of the bigger picture?

8. Rank, in your opinion, the following planning-related reasons why some projects perform poorly. Use a scale from 1 to 10, with the strongest cause as 1 and the weakest cause as 10.

- Inadequate period of time and effort spent on planning
- Lack of ability on the part of the planner
- Lack of management support for planning
- Lack of end-user* support for the plan
- Lack of involvement of end-users* in the planning
- Lack of justification of the plan
- Inadequate or non-existent monitoring
- Disregarding the plan once developed
- Monitoring of project performance but no feedback for corrective action
- End-users* not being properly informed of their duties under the plan.

[* Planning end-users refer to those who are to use the plan, the so-called 'doers'.]

List planning-related reasons, other than given above, as to why some projects perform poorly, and rank these reasons as well.

What are the causes of these practices? The use of a cause-effect, Ishikawa, fishbone diagram or similar might be useful here.

What suggestions can you make in order to eliminate the causes of these practices?

9. Estimate the cost involved in the planning of a project that you are familiar with. Base your estimate on the number of planner-hours that you think are involved.

Estimate the cost of delay (avoidable) in the project's completion date by one day, one week and one month. Base your estimate on component costs such as holding charges (interest on borrowed money), liquidated damages, overheads, equipment and facility hire, and wages.

Compare the two amounts estimated.

Can the degree of planning observed be justified on an economic basis?

In general, should planning effort be commensurate with the results required?

10. List possible sources of control that could be used in the following situations in order that the project does not diverge further from the planned state. No change in scope is allowed. (In particular, the specification cannot be altered.) No unethical practices are allowed. It is assumed

that the original planning and design for the project have been undertaken in an optimum fashion, and there is no 'fat' available for trimming.

- The project is running behind production/schedule.
- The project is going over budget.

11. Project duration and cost are related. But how are they related? There is a view put forward on large projects that 'Extra duration costs extra money. Reduced duration saves money.' What is your view?

12. The sale of something can be treated like a project. Imagine you are selling your car or house. What kind of planning would you undertake?

13. Plot, on the following diagram, your view of planning-related tasks over the duration of projects, from the point of view of:

- The owner.
- The main contractor.

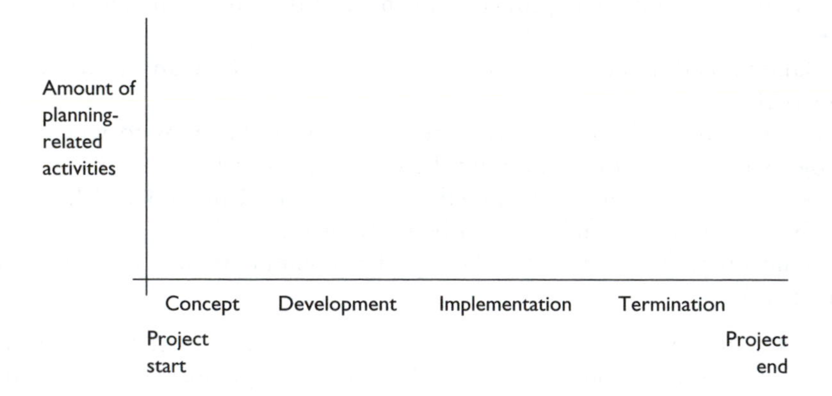

14. Both planning and design are inverse systems problems. Broadly, in both problems you are trying to determine the input of a system in order to achieve a desired output. This is in contrast to analysis where you determine the output of a system resulting from a given input.

List the conceptual similarities between planning and design.

15. Project close-outs commonly have a heavy emphasis on a fixed completion date. Planning may be done backwards from this point. What special planning problems would this create compared to planning being done in a forward sense?

Chapter 2

Systems thinking

Introduction

Framework

Much of this chapter is based on Carmichael (2004). The conceptual framework offered by control systems theory is regarded as the most suitable for developing planning thinking.

Thinking on planning is developed in two ways (Figure 2.1):

- The modelling of projects as control systems, both single level and multilevel, with attendant project descriptors of control, state and output variables.
- The interpretation of planning as an *inverse systems problem*, and in particular a *synthesis problem*. With project changes and with project evolution, replanning becomes an evolving set of similar inverse problems. With the evolution of a project, the inverse problem becomes better and better delineated. The same type of problem is being solved repeatedly throughout the project. Inverse problems by their nature have non-unique solutions; uniqueness is established within an optimal control systems formulation.

Projects in the following have been stripped-back or simplified to their mechanical backbone (Figure 2.2). Issues relating to people behaviour, which would ordinarily be superimposed on this backbone, are considered separately.

Systems–subsystems

If a project is regarded as a system, then just as a system can be broken down into subsystems, so a project can be broken down into subprojects. The decomposition may be continued to sub-subsystem (sub-subproject) level and beyond. (The process can also go in the other direction, from projects to

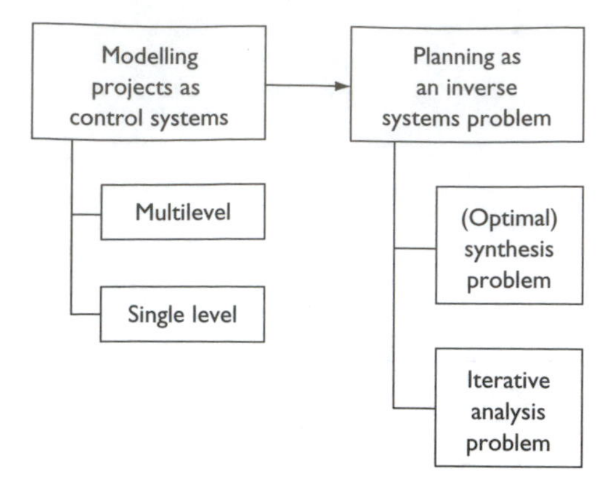

Figure 2.1 Idea development within this book.

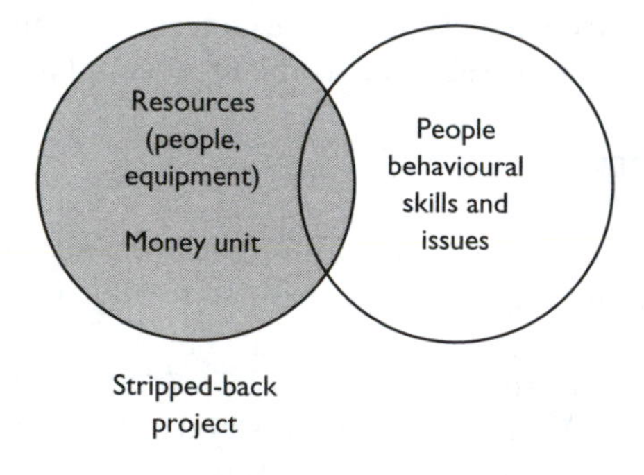

Figure 2.2 Shaded portion showing stripped-back project issues considered in this book.

'program', a 'program' being regarded as a collection of projects.) Decomposition can be carried out in a number of ways, for example Figure 2.3. The choice of decomposition will suit the intended project usage. In Figure 2.3, for 'system' read 'project', in terms of later chapter developments.

With the example decomposition of Figure 2.3a, the subsystems are all characterised similarly. With hierarchical decomposition of Figure 2.3b, successive decomposition leads to finer detail, and a change in the information type; the terms 'higher' and 'lower' are used when referring to levels in such a decomposition. (Figure 2.3a represents a two-level hierarchy even though it is called a staged (phased) decomposition.)

Commonly, project decomposition may be according to:

- Phase (chronological); time period phasing
- Work type or nature. Resource groupings
- Work parcels

- Contracts
- Region, location or geographical area
- Organisational breakdown
- (Existing) cost codes, cost centres, cost headings
- The nature of the planning network. Activity groupings
- Function

or generally according to:

- Project
- Subproject
- Sub-subproject, work package, job, . . .
- Activity
- Element
- Constituent.

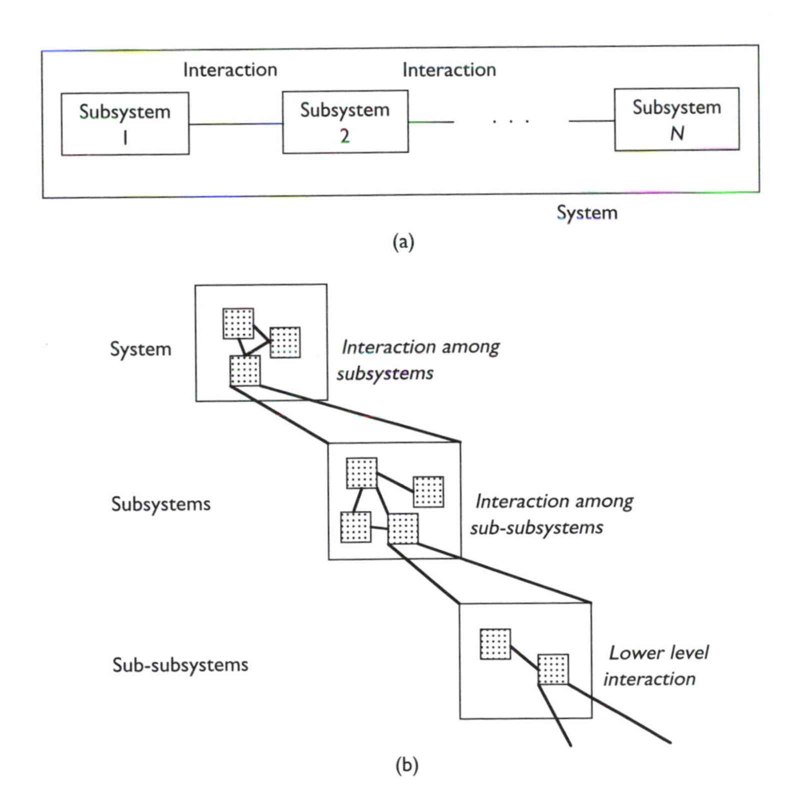

Figure 2.3 Example system decompositions (after Carmichael, 2004). (a) Staged (phased) decomposition; (b) Hierarchical decomposition.

Subsystems combine through their interaction to give the system at the next higher level.

Subsystems, sub-subsystems, . . . share the same properties and descriptors as systems, and hence are 'systems' in their own right.

Single level system

To understand systems ideas, it is convenient to start with a single level system. Multilevel, hierarchical systems (including staging) are collections of single level interacting subsystems.

Environment

The environment is defined as everything except the system (Figure 2.4). The environment affects the system by changes, and is affected by system changes. Constraints, project initial and final (terminal) conditions, and disturbance reflect the environment and the environment–system interaction.

Input and output

A system may be regarded as an input–output pair or input–output transformation (Figure 2.5). Input and output contain multiple parts (that is, they are vector quantities of input variables and output variables respectively), and the singular form of the terms is used to refer to both a single entity or plural entities, though sometimes the plural terms 'inputs' and 'outputs' may be used when the plural is specifically intended.

Subsystems and lower level components also have inputs and outputs.

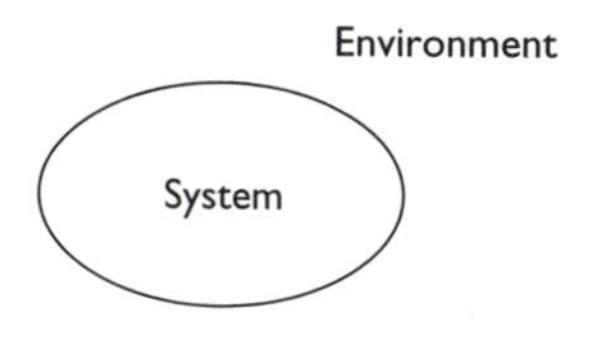

Figure 2.4 System, environment distinction.

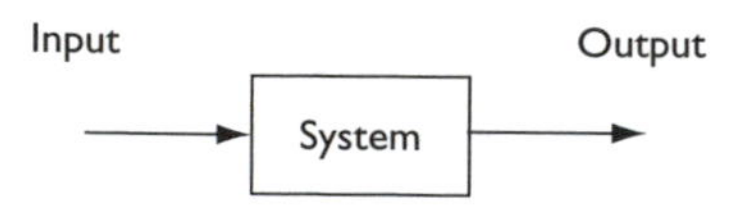

Figure 2.5 System as an input–output transformation.

Input, control

Input is equivalent to a decision, action or control. The term 'control' is favoured here in place of input, decision and action. The planner (decision maker) selects the project control in order to get a desired project output (performance, behaviour or response).

The preferred control is that which extremises the project objectives.

Control contains multiple parts (that is, it is a vector quantity of control variables), and the singular form of the term is used to refer to both a single entity or plural entities, though sometimes the plural term 'controls' may be used when the plural is specifically intended.

Output, state

The output describes the external or observable system performance, behaviour or response. (The terms 'performance', 'behaviour' and 'response' are used interchangeably in this book.) Internal system behaviour is usually given in terms of system state. For the stripped-back projects considered in this book, output and state are the same, that is there are no observability (in the sense of Kalman) issues.

Output and state are controlled variables. In planning, it is the project performance or behaviour which is being influenced by the choice of the control.

'State' contains multiple parts (that is, it is a vector quantity of state variables), and the singular form of the term is used to refer to both a single entity or plural entities, though sometimes the plural term 'states' may be used when the plural is specifically intended.

Disturbance

In most systems studies, there is something that prevents the hoped-for output being obtained exactly. This corruption is represented by disturbance in Figure 2.6.

Disturbance reflects influences (on project performance) that are different to that anticipated when (re)planning (and which prevent as-planned performance being attained). This includes the impact of unforeseen influences

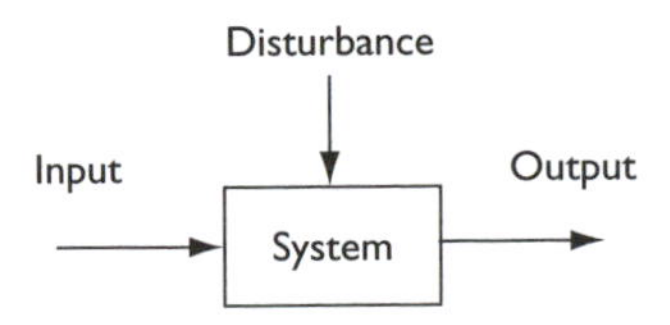

Figure 2.6 System with disturbance.

and influences that do not remain constant throughout a project. It implies uncertainty.

Automatic control – Technical systems

The automatic control of technical (mechanical, electrical, chemical, . . .) systems may be done in a number of ways. Of relevance to planning and replanning are:

- *Open loop control.* Information flow is one way. The control is selected up front, and no follow-up monitoring of system performance and corrective control or action is carried out (Figure 2.7a).
- *Closed loop (feedback) control.* There is a circular transmission of information (Figure 2.7b). Closed loop control is commonly used in an *error control* form. The control is selected and feedback occurs based on the difference between desired system performance and actual system performance.

Closed loop (feedback) control attempts to nullify the effects of disturbance on system performance, and to keep the system performance as desired. It has the better chance of ensuring that the system's performance goes close to that desired, compared with open loop control where the project performance may or may not go close to that desired.

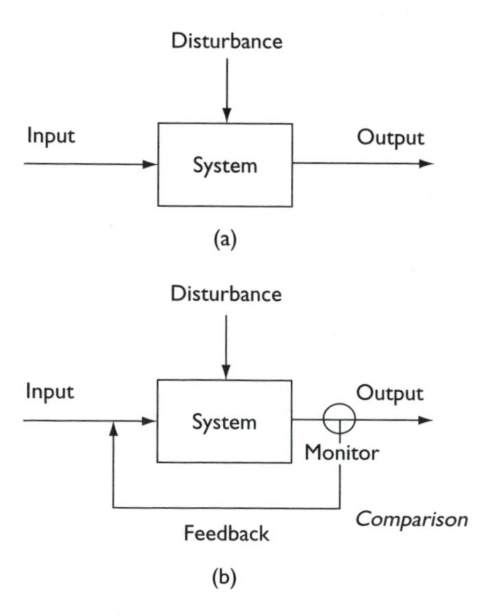

Figure 2.7 Comparison of open (a) and closed loop (b) representation.

In error control, feedback may be in a positive or negative sense. Negative feedback decreases the deviation between desired and actual performance, with the situation of zero deviation being the ultimate target. (Positive feedback increases the deviation.) There is no feedback if there is no deviation.

Both open loop and closed loop control include anticipation of the effect of future events/influences (much like the approach adopted in risk management). The control is selected with the view that future system performance is going to be affected by future events/influences.

The Deming PDCA (plan, do, check, act) cycle is no more than a closed loop control system.

Planning

Planning uses open loop control incorporating estimates of future influences. This gives a project 'baseline' or 'target' performance.

Replanning

Monitoring and reporting of project performance gives project status in terms of absolute performance and performance relative to the baseline (errors or variances).

Replanning uses open loop control incorporating updated estimates of future influences, and initial conditions equal to the current project state. Error control ideas are used to guide the choice of which controls need to be adjusted, but are not used in an automatic closed loop control sense in current replanning practice. Replanning gives an updated project baseline or target performance. Monitoring and reporting follow, and the cycle continues throughout the project.

Replanning modifications of control systems ideas include:

- Controls are not automatically selected, but rather the result of planner intervention. The planner adjusts the input (control) to a project (which may be subjected to future influences) to give a desired project performance.
- Many future events are not foreseeable or foreseen. This introduces (unwanted) deviations from desired performance, and this in turn can force a change in scope, work method, resource usage and resource production rates. Hence, for error control purposes, there is the possibility of the planned baseline of desired performance (from which errors are measured) changing as the project progresses.
- The measurement of project performance is done in most situations with not much accuracy, making precise control selections not possible. As well, the relationship between control and performance can be

complicated because of the innumerable things happening on a project at any one time.

- Without the instantaneous use of information coming from the monitoring of the project performance, time lags within the process interfere with the choice of effective controls. Collected project data may be overtaken by new events. Control is applied based on an earlier situation rather than the current situation.
- Both open loop and closed loop control are not used in a real-time sense.
- No state estimation is undertaken. For the stripped-back projects considered here, it is assumed that there is no noise in the observation device (though this is not true where people are involved in project monitoring/data collection). In the sense of Kalman, there is no observability issue. The output equals the state. There is also no controllability issue.

Fundamental systems problems

In terms of Figure 2.5, the fundamental systems problems are:

- *Analysis* – given the control/input and a model of the system, evaluate the output/response.
- *Synthesis* – given a model of the system and the output/response, evaluate the control/input.
- *Investigation* – given the control/input and the output/response, evaluate the model of the system.

In each, something different is known and it remains to evaluate the unknown.

Synthesis and investigation are known as *inverse problems*. In general the solution of such problems is *non-unique*, whereas the solution of the analysis problem is generally unique. Non-uniqueness in the synthesis and investigation problems might be removed through additionally including some objectives or equivalent. (Note that the presence of more than one objective introduces subjectivity, which is not present in the single objective case.)

Planning and design (and management) are seen to fit within the synthesis categorisation.

Synthesis requires the a priori specification of a systems model, that is the relationship between input and output.

Synthesis problem components

The synthesis problem is essentially a converse to the analysis problem. Given a certain (desired) response and model, the issue is to determine

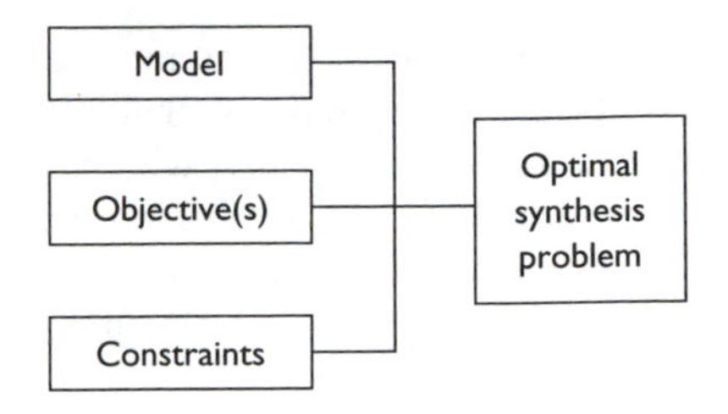

Figure 2.8 Synthesis problem components.

the control that produces this response. Generally, a certain response is realisable with many control choices. That is, the solution is *non-unique*. Further requirements such as extremisation (optimisation, maximisation or minimisation) of objectives, for example minimum duration or minimum cost, are needed to make the solution unique.

See Carmichael (2004) for a rigorous meaning of the term 'objective'. Generally one desires to synthesise a system which is optimal in a certain sense; objectives (or equivalent terms) resulting from an imposed value statement are implied.

The (optimal) synthesis problem is seen to have three main components (Figure 2.8) (Carmichael, 1981):

1. Model
2. Objective(s)
3. Constraints.

These problem components are expressed in terms of control, state and output variables.

Optimal synthesis

The optimum problem is within the realm of the well-delineated body of theory and techniques of optimal control systems or optimisation theory. In this sense the planning problem is a single level or multilevel, single or multiple objective optimal control problem.

Optimal control theory is found extremely useful in elucidating the planning problem. Using state and control variable ideas, the planning problem can be readily exposed and understood.

In simple terms, synthesis is equivalent to choosing the project control; optimal synthesis or optimal control selects the control so as to extremise some objectives. In addition, supplementary constraints are also usually present.

Optimal control theory has applicability across all technology and management.

Conversion of synthesis to an iterative analysis problem

The solution of the synthesis problem could be expected to be more complicated and more difficult than the solution to the analysis problem. In some cases the analysis problem can even be solved intuitively, whereas the solution of the synthesis problem involves more rigorous methodology.

Because of the degree of difficulty of solving synthesis problems compared to analysis problems, many synthesis problems are solved in an iterative analysis fashion.

Established planning procedures are iterative in nature. The iterations arise from the analysis-based mode of attack on the planning problem and are not inherent in planning. That is, established planning practices use iterative analysis.

The writer's preferred analysis mode of attack on problems is through the body of knowledge on systematic problem solving, or equivalently systems engineering.

Systematic problem solving goes through the following steps (together with iterative or feedback loops):

- Defining the problem (including issues and situation)
- Selecting objectives
- Generating ideas, alternatives
- Analysing ideas, alternatives
- Selecting the best alternatives (evaluation)
- Action.

Alternatively, a systems engineering problem goes through the following steps (together with iterative or feedback loops):

- Defining the problem (including issues and situation)
- Selecting objectives
- Synthesising systems
- Analysing systems
- Selecting the best alternatives (evaluation)
- Action.

See Carmichael (2004).

Feedback occurs between steps in trying to refine the problem (Figure 2.9). A systematic approach to the above steps generates clear thinking at each step and lends objectivity to a procedure which would otherwise be considered intuitive.

The objectives are evaluated for each set of values of the control variables. Similarly the performance, behaviour or response is evaluated for each set of these values. It is hoped that by adjusting the initial guesses for the control variables, the response resulting from the analysis will become more

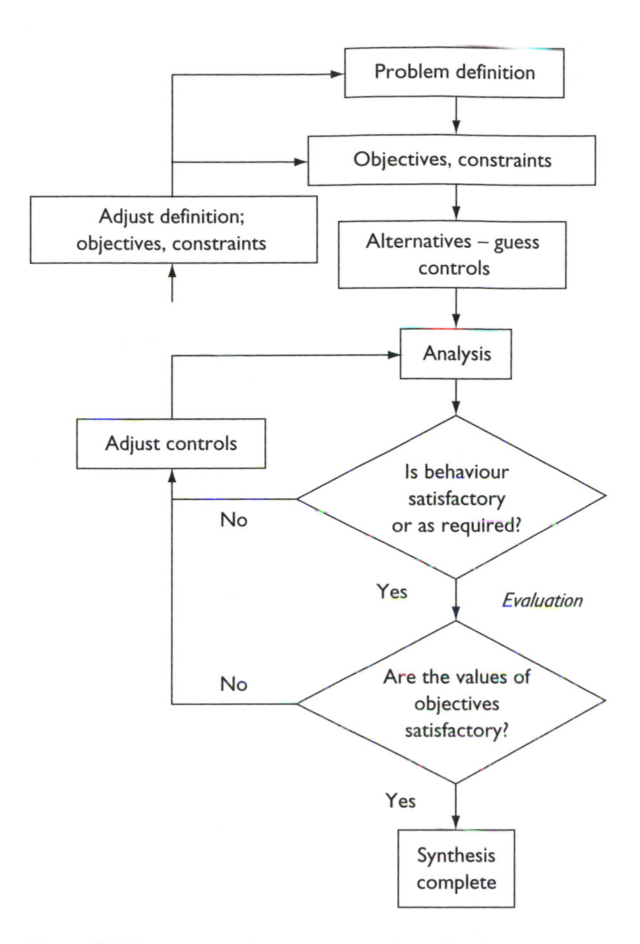

Figure 2.9 Iterative analysis version of synthesis.

favourable and also the values of the objectives will improve (decrease or increase as appropriate). However there is no guarantee of this, although the experience and knowledge of the person attacking the synthesis problem will usually head the iterations in the preferred direction. Where this person has synthesised such systems before, the person's first guess of values for control variables may be close to 'optimum' and no iterations may be required; nevertheless the synthesis is still being approached via iterative analysis. Established, experienced planners would be expected to make good first guesses and thereby eliminate the iterations. Beginner planners, on the other hand, would be expected to do several iterations.

The end point of the iterative modification with feedback loops is based upon obtaining satisfactory project performance and satisfactory values for the project objectives.

Chapter 3

Planning tools and terminology A–Z

Activity

Also called task. Technical disciplines tend to use the term 'activity', while non-technical disciplines seem to prefer the term 'task'. The term 'activity' is used in this book.

The term 'activity' may refer to a physical activity, an administration or a management activity, a material delivery activity, or anything that has a finite duration and uses resources (and consequently money); an identifiable piece of work.

Exceptions to this include:

- A *dummy activity* which exists to maintain work method logic in a network, has zero duration and uses no resources. A dummy activity might be referred to as a logic restraint. Project start and project end activities may also be optionally introduced in order to give identifiable start and end points to a project.
- An *artificial activity* which has a finite duration but uses no resources. An artificial activity may be used, for example, where a schedule delay or constraint is required. Occasionally, certain activities have to be completed by a prescribed date or within a prescribed time period (a duration limitation). This constraint may be straightforwardly handled by introducing artificial activities into a network, where the duration of the artificial activity equals that prescribed time period.
- An *aggregated activity*, which is a collection or group of activities, used to give economy of presentation where detail is not required by the viewer. The terms 'summary activity' or 'hammock' may also be used.

An activity occurs at the lowest level of a work breakdown structure; the level at which detailed duration and resource estimates (and consequently money estimates) are commonly applied.

Activities may be summarised for end-user purpose according to:

- Description
- Associated work method; incorporated materials (including potential suppliers, origin and freight)
- Logic sequence (relationship to other activities)
- Specification and contractual related matters
- Resource needs
  - People (engineering and other services; personnel and organisational responsibilities)
  - Equipment (including facilities, manufacturing and origin)
- Activity production rate
- Duration
- Money (cost and income) estimate. Cost account number
- Assumptions.

A *continuous* activity is one that once started continues until complete. An *intermittent* activity is one that can be stopped and started with breaks in between.

Bar chart

Also called Gantt chart and Gantt bar chart (after Henry L. Gantt's work in the early 1900s). Technical disciplines tend to use the term 'bar chart', while non-technical disciplines seem to prefer the term 'Gantt chart'. The term 'bar chart' is used in this book.

A bar chart may be referred to as a *program* or a *schedule*. But there are other diagrams and charts which are also referred to as programs and schedules.

A bar chart contains information at both the activity and the project levels. A bar chart has a timescale horizontally (but no scale vertically), with the activities listed vertically. The horizontal scale is ideally matched to a calendar allowing for stipulated work hours per day, weekends, public holidays, rostered days off (RDO), number of shifts per day and so on. The activity listing may be in any order, though usual preference (to facilitate project monitoring) and common practice is to list activities in order of their earliest start times, leading to a stepped appearance from the top left corner to the bottom right corner. Alternatively, like activities may be grouped. A horizontal bar is drawn adjacent to each activity's name showing that activity's scheduled duration, between the start and end times of that activity, and float may also be added. Bar charts may be subdivided (horizontal partitions) into subprojects. Activities might be collected together (as also in a subproject) and represented by a single aggregated activity.

Activities are preferably shown connected, as in Figure 3.1, indicating the order or logic of work or activity dependence. Such a diagram may be referred to as a *connected* (or linked) *bar chart*. In such a form, a bar chart is an activity-on-link diagram drawn to a timescale. It assists both planning and replanning; the influence on following activities of an activity change can be readily established.

In Figure 3.1, the scheduled duration of an activity is shown by a hollow box; float is indicated by a shaded box. Activities are preferably shown with float.

Although a bar chart can be drawn directly by scribbling horizontal lines on a page, a more rational way of obtaining a bar chart is *via a network (critical path) analysis* because this includes the work method and float information. The practice of drawing lines on a page (without going to the trouble of a critical path analysis) to give the impression of a rationally developed bar chart has little merit. The information from a network analysis is readily transferable into bar chart form. Activity start

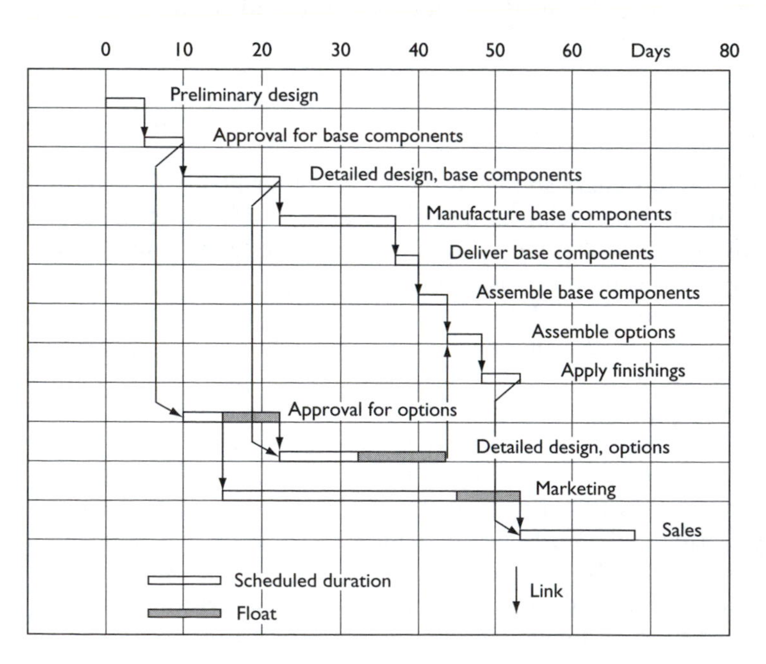

Figure 3.1 Connected bar chart for an example project. See the associated networks, Figure 3.25(a,b), and time-scaled network, Figure 3.55. (Activity names have been abbreviated because of space limitations – work items are implied.)

and finish times and floats are calculated in the network analysis. The connections between activities reflect the logic of the network diagram. (Some computer packages might appear to go in the reverse direction [that is, bar chart to network] but this is not the case – the so-called bar chart developed first is in fact a time-scaled [activity-on-link] network.)

Figure 3.1 shows the bar chart plotted for all activities starting at their earliest start times. However, the start times of the non-critical activities can be varied because of their float. Hence, in principle, a large number of bar charts are possible for any given project. This flexibility is used by planners to optimise the usage of resources.

In summary, a bar chart gives information on when activities begin, when they end, how long they take, the amount of float, and activity dependence with other activities.

With connections between activities, the bar chart contains the same information as a network, but usually is more readable. As noted, a connected bar chart is a time-scaled activity-on-link network. When 'integrated', a bar chart becomes the same as a time-based cumulative production plot. Cumulative production plots, bar charts and multiple activity charts contain essentially the same information.

Bar charts also find use in reporting for replanning purposes, where actual project performance is included additional to planned performance.

Baseline

A baseline refers to the as-planned situation. It may be referred to as a *target*.

The baseline may be with respect to resource usage (as-planned resource plot, as-planned cumulative resource plot), money usage (budget, as-planned cash flow), schedule (as-planned program), production (as-planned cumulative production plot), and delays (as-planned cumulative delay plot).

Actual or as-executed progress and performance is compared with the various baselines for resource types, cost, production and so on, as part of reporting.

Budget, budgeting

Refer 'financial planning'.

Cash flow

The difference between income (*inflow*) (payments received) and expenditure (*outflow*) (payments to suppliers, employees, trade contractors, interest on own or borrowed funds, . . .) is the cash flow, which can be shown in a cash flow diagram or a table. A *positive cash flow* indicates that more money has been received than has been paid out; a *negative cash flow* indicates the opposite situation.

Closed loop control

Closed loop control may be referred to as *feedback control.* It is commonly used in an *error control* form; the control is selected based on the difference between desired system behaviour and actual system behaviour.

Compression

Project compression refers to the reduction or shortening of a project's duration. The project compression problem is a subproblem of the total planning problem, where emphasis is placed on the different level controls that can lead to activity and project duration shortening, and all other controls are held fixed.

As part of an iterative analysis attack on planning, project compression may be considered:

- Where the calculated project duration, based on normal (least cost) durations, is excessive.
- In order that the overall project completion time meets some required completion time or deadline, perhaps where penalties or damages for completion time overrun start to accrue. A similar situation arises should there be bonuses for early completion.
- When, part way through a project, the project is running behind schedule. Material, equipment or worker/personnel shortages may develop during a project thereby delaying the project completion time. Delays due to other causes (such as those due to weather, industrial disputes, unforeseen workplace conditions, . . .) may occur during project execution.
- To assist resource and money allocation on a project or over several projects, it may be desirable to complete a project early in order to, say, reuse the equipment or workers/personnel on another project.
- Where there may be large time varying indirect costs, decreasing the project duration may give a lower total (direct plus indirect) cost solution. See 'estimates' for a distinction between direct and indirect costs.

A project's duration may be reduced by reducing or shortening the durations of critical activities (*activity compression*).

Compression may come at a cost penalty.

Reducing the durations of non-critical activities does not reduce a project's duration, but introduces more float, and might be considered as a possible way of utilising resources more effectively.

In the compression of activities, attention focuses on what are termed *normal durations* and *costs*, and *crash durations* and *costs*. Activity normal durations are those of least direct cost and are usually the durations for which the initial planning calculations are done. Durations cannot go below activity crash durations – they are the absolute minimum activity durations obtained by employing maximum resource numbers (Figure 3.2). *Crashing* consequently refers to taking compression to the limit. A theoretical construct called a cost slope can be used to rank activities and establish an order for compressing. A cost slope may be defined as the relative increase in cost per unit time saved undertaking an activity. Activities with the lower *cost slopes* are chosen first for compressing because these lead to lower increases in total cost. The order of compression is selective. The use of such an approach in practice will depend on having adequate estimating data, and this may not be available.

Figure 3.2 represents the conceptual form of activity cost–duration data. Generally, however, the continuous range of cost–duration information shown in the curve is not available to planners. Instead, estimating, manufacturers' or contractors' information may be limited to only two or three cost–duration data points as shown in Figure 3.3. The number of possible ways of undertaking an activity may also reduce the data to a few (finite number of) points. Cost–duration data points represent the minimum costs to perform the activity in the given durations. In such cases, where it is only possible to undertake an activity in a finite number of ways, for example by using a team of three workers or a team of five workers, cost–duration data points referring to situations in between these defined points are meaningless. As another example, consider a delivery activity; the item or material might be delivered by air within a short time period, but high cost, or by

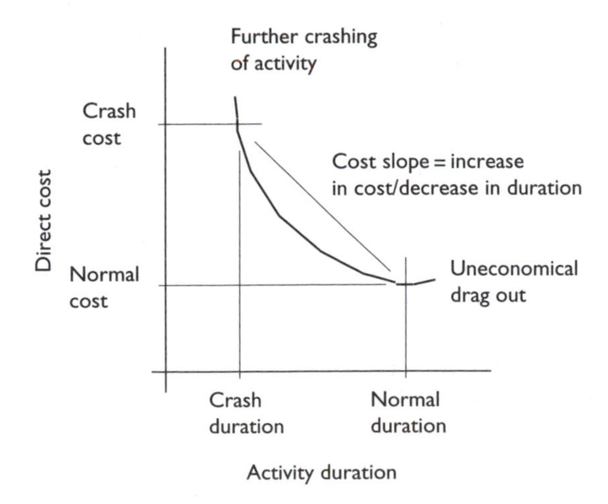

Figure 3.2 Compression terms.

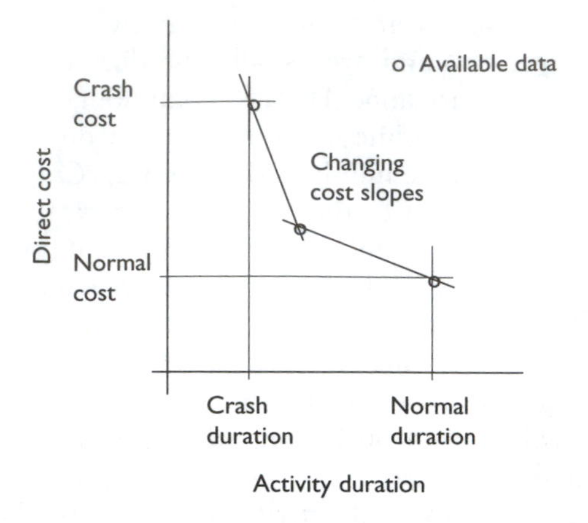

Figure 3.3 Available activity cost–duration data.

sea at less cost but longer time period. These two modes of delivery define two cost–duration points. The interval between these two points is not defined. Any assumption relating to a continuous cost–duration trade-off is not relevant.

However, the treatment of this *discrete case* is similar to that in which a continuum of possible cost–duration data points exists (*continuous case*). A special form of the discrete case is when the normal and crash points are the same; that is the cost–duration data is confined to one point and there is only one way of undertaking that activity.

A variant of the cost–duration diagram of Figure 3.3 occurs when there are two or more distinct ways of carrying out an activity. Figure 3.4 indicates the idea where there are two methods available, and hence two distinct cost slopes depending on the method used.

A number of writers raise the issue of the validity of assumptions relating to continuity and linearity of cost–duration curves. A possibly more serious issue is the (un)availability of multiple cost–duration data for any activity, and the accuracy of this data.

Compression without activity cost data can also be carried out, but here the choice of which critical activities to compress becomes almost arbitrary. There are many ways that activities can be selected.

Project compression may be viable because, even though the direct costs of activities increase and hence the direct cost of the project increases as the project duration is shortened, time varying indirect costs are less for shorter duration projects. There is consequently a trade-off between increased direct cost and reduced indirect cost (Figure 3.5). If there are large time varying indirect costs, decreasing the project duration may give a lower total (direct plus indirect) cost solution.

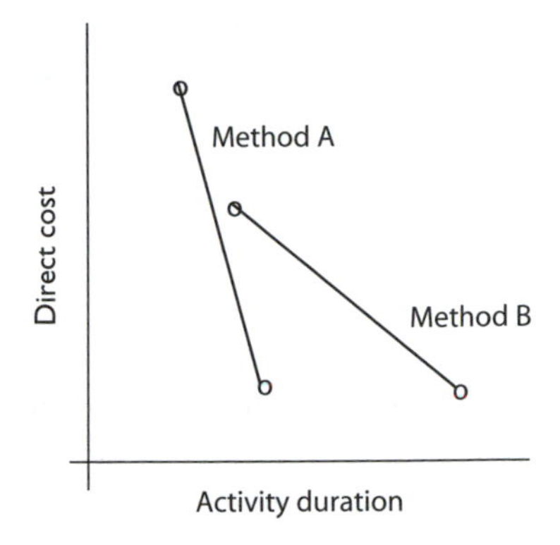

Figure 3.4 Example cost–duration data for two work methods.

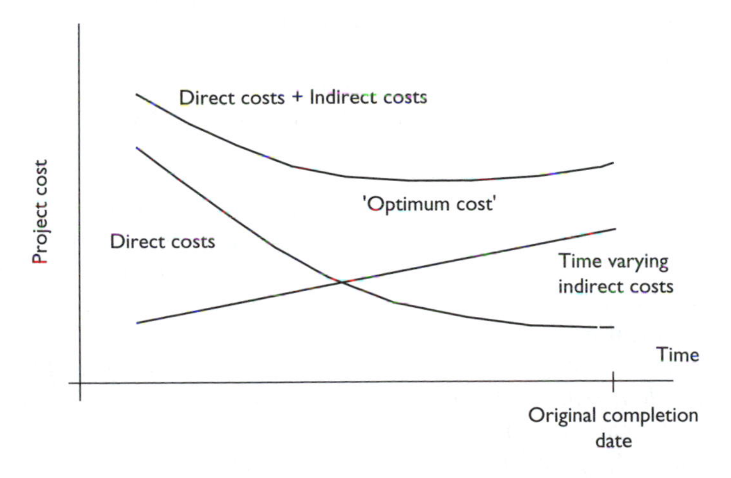

Figure 3.5 Schematic of direct cost and indirect cost trade-off involved in project compression.

For control selection, in the compression problem, general ways of shortening a project that fit within the above discussion include:

Constituent level

- Change the resource production (constitutive) relationship
- Use a carrot (incentive, reward)/stick approach.

Element level

- Use additional resource numbers (quantity)
- Work overtime or multiple shifts.

Activity level

- Trade-off additional resource numbers (quantity) versus activity duration.

Project level (changing method)

- Run concurrent/parallel activities
- Use overlapping relationships.

The solution steps to the project compression subproblem are no different to any project planning problem.

Compression algorithm

Adopting an iterative analysis approach to planning, it is possible by examining a network and the associated cost–duration relationships for the activities, to manually compress a network, or to write a computer program that mimics the manual procedures. Coarse steps in the process follow Figure 3.6.

The compression calculations may pass through several phases. The three most notable are outlined below:

- Compressing of critical *activities may proceed to their respective crash durations* and no lower. If the status of other activities has not been affected during this process, then the compression calculations are complete. (Similar comments apply to the simultaneous compressing of multiple critical paths.)
- Compressing of critical activities may proceed *until a non-critical activity has its float reduced to zero* and hence itself becomes critical (and hence itself a candidate for compressing).
- Where *multiple (parallel) critical paths* exist in a network, then any compressing of activities has to be carried out simultaneously and by the same time interval on all critical paths. Compressing on less than the full number of critical paths will not reduce the total project duration.

When no further compression is possible, the process is complete.

Some oddities that may occur in connection with network compression are as follows. Often compressing of one activity which leads to overtime being worked on that activity may lead to other activities also being carried out on overtime purely to maintain industrial harmony in the workplace even though the project may not require these other activities to be carried out in other than normal durations. There is also the possibility that in order to compress one activity it may be necessary to introduce a new piece of equipment to the project. This piece of equipment may then be used on other activities to gain maximum use out of it although, again, these other activities may not require speeding up.

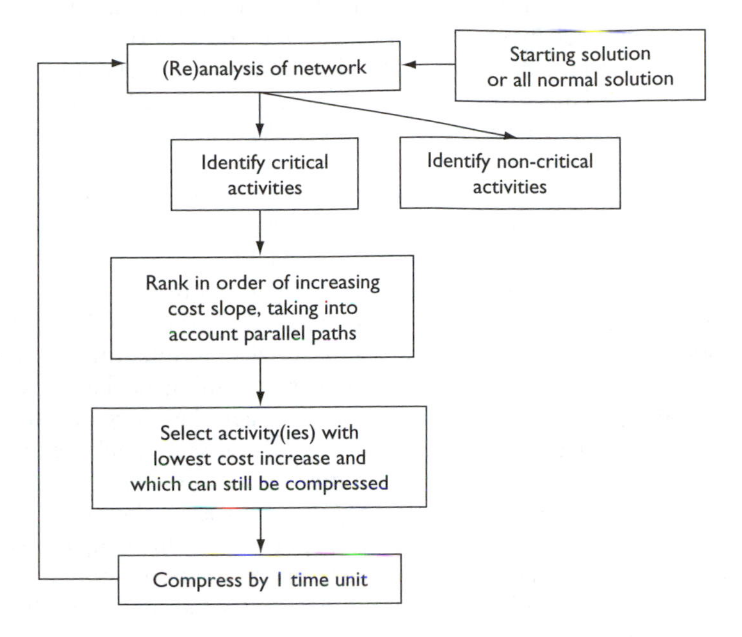

Figure 3.6 Steps in project compression.

Computer packages

There are many computer packages of varying user-friendliness available on the market to assist project planning. They perform the analysis part of the iterative analysis mode of attack on planning. They may advertise 'control' capabilities, but this is 'control' in a lay person's loose meaning of the word. They treat planning subproblems such as resource smoothing, resource constrained scheduling and project compression, but in replanning, at best they only provide information on which replanning is based.

Computer packages based on the critical path method of analysing networks have been around for several decades now. The early packages operating on mainframe computers required punched card input and batch processing with slow turnaround time period between computer runs. This was replaced by mini computers and paper tape input. The latest computer packages exploit the graphics capabilities of personal computers, and involve editing directly on the monitor and almost instantaneous results for the user. (However, many established planners still prefer to draft the project network by hand on paper before transferring it to the monitor.) The latest packages are also considerably more sophisticated in the number of options they allow and their computer input and computer output capabilities.

Several decades ago, it was not uncommon for an organisation to have a computer program especially written for scheduling and reporting. All programs were based on the critical path method. Journals of that era published listings of simple computer programs. There arose a number of commercial packages from this background, generalisations of organisations' in-house packages, in an attempt to tap a larger market and get a return on the invested resources writing the computer programs. Some packages became well known, but all of that era seem to have disappeared and have been replaced or upgraded with more user-friendly packages, in line with the more user-friendly graphical interfaces of modern computers.

Today, there are many packages in the marketplace, varying in their price and user-options. Both activity-on-node and activity-on-link formats are catered for. Organisations also supplement these planning packages with word processing, spreadsheet and database packages. Few organisations write their own software any more. Most use off-the-shelf packages and adjust their internal procedures to the limitations this imposes.

Frequently, the choice of package that an organisation uses is dictated by the package requested by the client, and so some adaptability of thinking between packages is required. Fortunately, the packages tend to have common bases.

Planners need to be careful not to fall into the trap of accepting without questioning anything that comes out of a computer analysis. There is the acronym GIGO (garbage in, garbage out) and of course Murphy's Law (extended) has something to say on the matter: *To err is human, but to really foul things up requires a computer.*

A suggested approach is to always interact with any computer package, leaving the routine analysis calculations to the computer package but not the decision making (control selection).

Constituent

The basic level in a project hierarchy, referring to resource behaviour.

Constraints

The plural terms 'objectives' and 'constraints' are generally used in this book even though the singular forms may be applicable in some situations; the exception is where the singular is deliberately intended.

All projects have genuine constraints such as funding, environmental (natural) and political constraints. Constraints restrict the range of control choices possible.

As with objectives, many people confuse a project constraint with an end-product constraint. End-product constraints may influence project constraints.

Constraints may be stated at all project levels.

Refer Carmichael (2004).

Contingency

Refer 'reserve'.

Control

Planning establishes the value of control throughout the project duration. Control has multiple parts (that is, it is a vector quantity of control variables) that contain information on method (including sequence), resources (and hence money) and resource production rates (or equivalent).

The term 'control' is favoured in this book over input, decision and action. The planner selects the project control in order to get a desired project output (response, performance, behaviour). The preferred control is that which extremises the project objectives.

Critical activity, critical path

A *critical activity* is usually defined as one with no float. A *non-critical activity* is one with float. The *critical path* is the set of network-connected critical activities from project start to project end determining the project duration; it is the longest duration path through a network. If the start or finish of a critical activity is delayed, the project duration will be extended.

It is possible to have multiple critical paths in a network.

A critical activity is akin to a Pareto item, amongst the collection of all activities, though an 80:20 distribution of non-critical activities to critical activities rarely occurs. It is recommended that most attention be paid to the critical activities during project implementation in order that the project does not run over schedule.

The terms 'critical' and 'critical path' are popularly used terms in the project management field, often used incorrectly. Network analysis gives definiteness to these ideas.

Critical path method, CPM

The critical path method (CPM) is based on a deterministic *network analysis*, and broadly is taken to include all things connected with such an analysis. The term 'critical path analysis' (CPA) is also used. The analysis relies on a single duration estimate for each activity.

The network analysis takes advantage of the special directed characteristics of the network. In particular, the analysis is done in essentially three stages.

The first stage is a *forward pass* through the network in which earliest times (or dates) are calculated for each activity, namely:

Earliest start time (EST)
Earliest finish time (EFT)

The second stage is a *backward pass* through the network in which latest times (or dates) are calculated for each activity, namely:

Latest start time (LST)
Latest finish time (LFT)

Thereafter (the third stage) the total float (TF), free float (FF) and interfering float (IF) may be calculated.

F/S relationship calculations

Mathematically, the calculations are as follows for activities with finish-to-start (F/S) relationships (and zero lead time period) between them. Similar calculations are performed where other relationships (that is, start-to-start (S/S), finish-to-finish (F/F), start-to-finish (S/F) and finish-to-start (F/S) – all with finite lead time periods) apply.

Forward pass

Initial conditions for the first activity(ies):

$$\mathrm{EST} = 0$$

$$\mathrm{EFT} = \mathrm{EST} + \text{duration}$$

For all subsequent activities:

$$\mathrm{EST} = \text{maximum EFT of preceding activities}$$

$$\mathrm{EFT} = \mathrm{EST} + \text{duration}$$

Backward pass

Terminal conditions for the last activity(ies):

$$\mathrm{LFT} = \mathrm{EFT}$$

$$\mathrm{LST} = \mathrm{LFT} - \text{duration}$$

For all previous activities:

$$\mathrm{LFT} = \text{minimum LST of following activities}$$

$$\mathrm{LST} = \mathrm{LFT} - \text{duration}$$

Float

For all activities:

$$\begin{aligned} TF &= LFT - EFT \\ &= LST - EST \end{aligned}$$

$$FF = \text{EST of following activity} - \text{EFT of activity itself}$$

$$IF = TF - FF$$

Setting the earliest start time of the first activity(ies) to zero is optional. Other values can be prescribed.

For the last activity(ies), setting the latest finish time equal to the earliest finish time means that the so-called critical activities will have zero total float. Terminal conditions other than the one mentioned above can be prescribed and this would be the case, for example, where there was a desired finish date for the project.

The project completion date is the EFT or LFT of the last activity(ies).

Overlapping relationships

The above calculations can be generalised for all overlapping relationship types. The notation of Figure 3.7 applies.

A full network analysis requires all network interconnections to be considered.

The calculations on networks involving overlapping relationships proceed in a similar fashion to the calculations of elementary network analysis. That is, on a forward pass, activity EST and EFT information is obtained and then on a backward pass, activity LFT and LST information is obtained. The floats and the critical path may then be determined. The relevant expressions are as follows for a network starting at activity 0 and finishing at activity N:

Forward pass calculations

$$EST_0 = 0$$

$$EFT_0 = EST_0 + DUR_0$$

$$EST_j = \max_{\text{all } i,j} \begin{cases} EFT_i + LT & F/S \\ EST_i + LT_i & S/S \\ EFT_i + LT_j - DUR_j & F/F \\ EST_i + LT_i + LT_j - DUR_j & S/F \end{cases}$$

The above implies that for more than one preceding activity, i, or overlapping dependence relationship choose the *maximum* EST_j value because the longest path through the network is sought. Where EST_j turns out to be negative, upgrade this value to 0.

$$EFT_j = EST_j + DUR_j$$

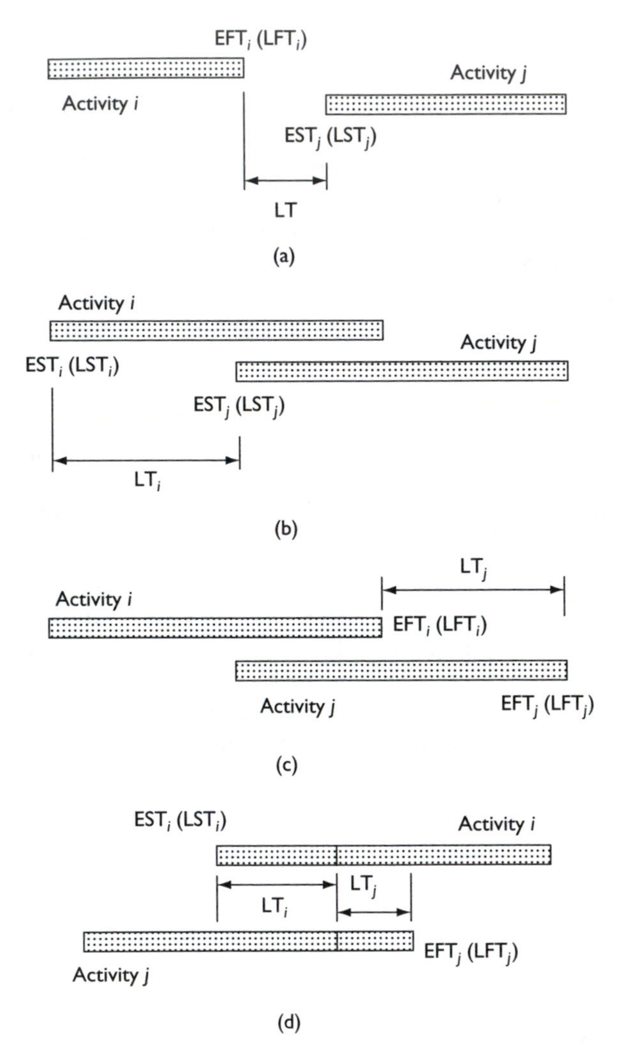

Figure 3.7 Lead time periods; two activities only. (a) Finish-to-start relationship; (b) Start-to-start relationship; (c) Finish-to-finish relationship; (d) Start-to-finish relationship.

Backward pass calculations

$$\mathrm{LFT}_N = \mathrm{EFT}_N$$

$$\mathrm{LST}_N = \mathrm{LFT}_N - \mathrm{DUR}_N$$

$$\mathrm{LFT}_i = \min_{\text{all } i,j} \begin{cases} \mathrm{LST}_j - \mathrm{LT} & \mathrm{F/S} \\ \mathrm{LST}_j - \mathrm{LT}_i + \mathrm{DUR}_i & \mathrm{S/S} \\ \mathrm{LFT}_j - \mathrm{LT}_j & \mathrm{F/F} \\ \mathrm{LFT}_j - \mathrm{LT}_i - \mathrm{LT}_j + \mathrm{DUR}_i & \mathrm{S/F} \end{cases}$$

The above implies that for more than one following activity, j, or overlapping dependence relationship choose the *minimum* LFT_i value.

$$\mathrm{LST}_i = \mathrm{LFT}_i - \mathrm{DUR}_i$$

Floats

Total float,

$$\mathrm{TF}_i = \mathrm{LFT}_i - \mathrm{EFT}_i$$
$$= \mathrm{LST}_i - \mathrm{EST}_i$$

TF_N will equal zero if the latest finish time is set equal to the earliest finish time for the last activity N. In such a case the activities with total float equal to zero will define the critical path. Where the latest finish time is not set equal to the earliest finish time for the last activity, TF_N will equal the difference between these two values and the critical path through the network will be defined by activities having total floats equal to this number. Note that where the largest activity LFT is not the network final activity, the activities at the end of the network (following the activity with the largest LFT) will be calculated to have a zero total float yet they are not critical activities. These activities can be allocated float in a post-analysis check.

Free float,

$$\mathrm{FF}_i = \min_{\text{all } i,j} \begin{cases} \mathrm{EST}_j - (\mathrm{EFT}_i + \mathrm{LT}) & \mathrm{F/S} \\ \mathrm{EST}_j - (\mathrm{EST}_i + \mathrm{LT}_i) & \mathrm{S/S} \\ \mathrm{EST}_j - (\mathrm{EFT}_i + \mathrm{LT}_j - \mathrm{DUR}_j) & \mathrm{F/F} \\ \mathrm{EST}_j - (\mathrm{EST}_i + \mathrm{LT}_i + \mathrm{LT}_j - \mathrm{DUR}_j) & \mathrm{S/F} \end{cases}$$

The above implies that for more than one following activity, j, or overlapping dependence relationship, choose the *minimum* FF_i value. These expressions represent the difference between the right- and left-hand sides of the earlier expressions for earliest times.

A note

For activity-on-node networks with non-overlapping relationships, redundant links occur with triangular logic patterns as shown in Figure 3.8.

Redundancy will similarly occur in networks with overlapping relationships provided the earliest start time of C (Figure 3.8) is determined by the path A–B–C; redundancy will not occur if the earliest start date is determined by the link A–C. Hence triangular patterns may or may not be permissible in networks with overlapping relationships. To check for redundancy the earliest start time of C has to be computed along the path A–B–C and along the link A–C. If the value on the link A–C is greater then remove the redundant link A–C.

Representation

Personal preference dictates the way that the activity start and finish times and floats are displayed on the network diagram. Figure 3.9 shows a number of examples. Each person, text and computer package has its own preference. Computer packages may allow customising the display.

The numerous commercial computer packages available to assist project planning, all have CPM as their basis.

To many people, network analysis is 'project management'. The reasoning goes that if you have a computer package, then you can set yourself up in business as a 'project manager'. This is not a realistic appraisal of project management.

Occasionally people (incorrectly) use the term 'PERT' (program evaluation and review technique) when talking about CPM. PERT is a probabilistic analysis method and appears not to be used by practitioners (even though the term PERT is used, but wrongly).

It has been humorously commented that: *The critical path method is a management technique for losing one's shirt under complete confidence.*

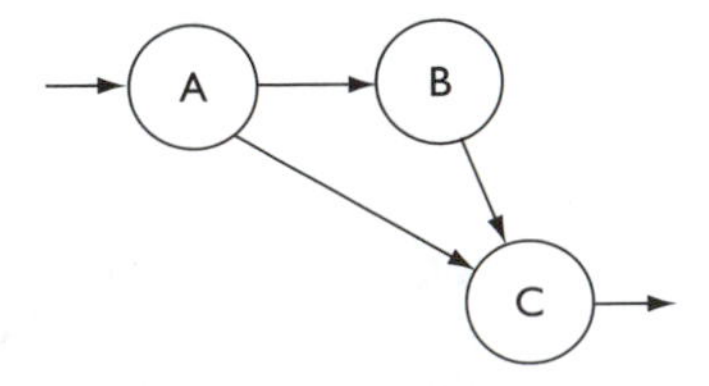

Figure 3.8 Redundant link example.

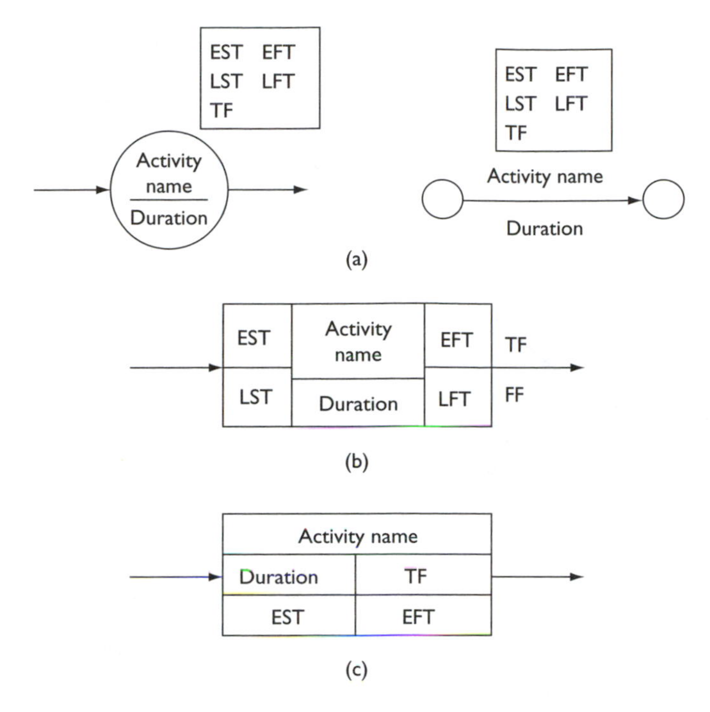

Figure 3.9 Example drafting styles.

Cumulative production plot

A cumulative production plot shows an activity's or project's production up to any time or distance (as the independent variable); the slope of the plot represents the rate of production (Figure 3.10). Figure 3.10 can be either an activity level or project level representation.

Some variants on Figure 3.10, at the activity level, are:

- The axis labelling may be interchanged, with the independent variable on the vertical axis (Figure 3.11).
- The diagram may be drawn down the page (Figure 3.11).
- The line plot may be given a finite width where work occupies durations of significance (Figure 3.12).

For a bar chart such as Figure 3.13a, joining the bars (that is, 'integrating' the bar chart) and changing the vertical axis gives the cumulative production plot of Figure 3.13b.

A cumulative production plot may be referred to as a *program* or a *schedule*, in some situations may alternatively be called a *trend plot*, *trend graph* or *time-chainage chart*, and is used in the line of balance technique.

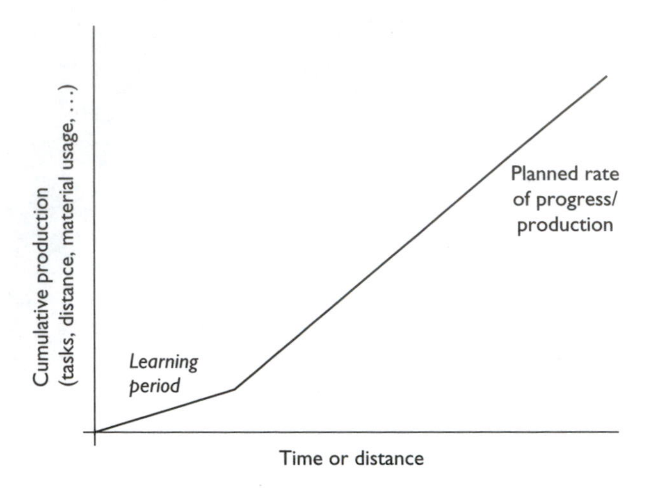

Figure 3.10 Example cumulative production plot.

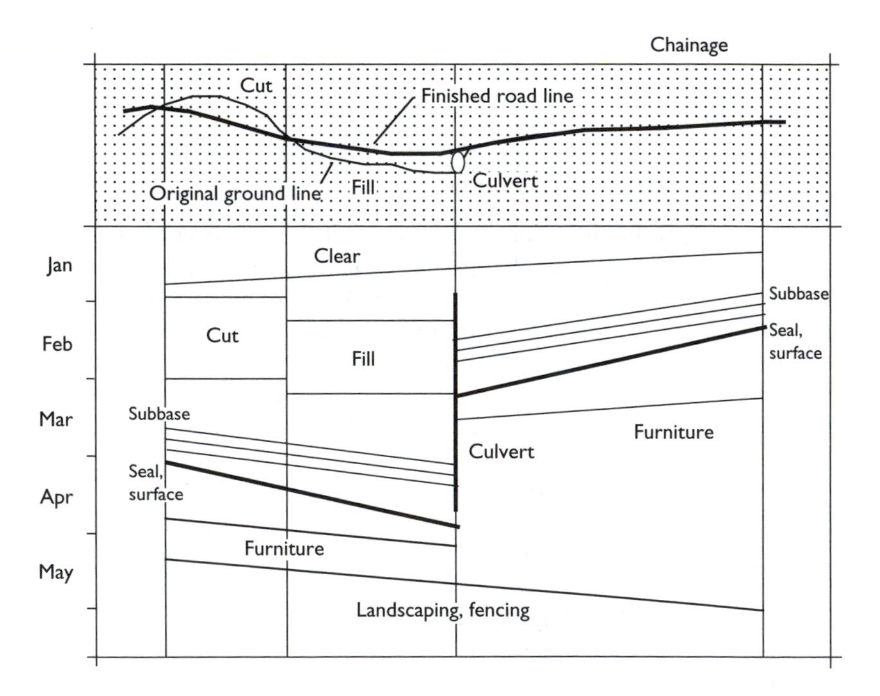

Figure 3.11 Example cumulative production plot (time-chainage chart) for road construction. (Activity names have been abbreviated because of space limitations – work items are implied.)

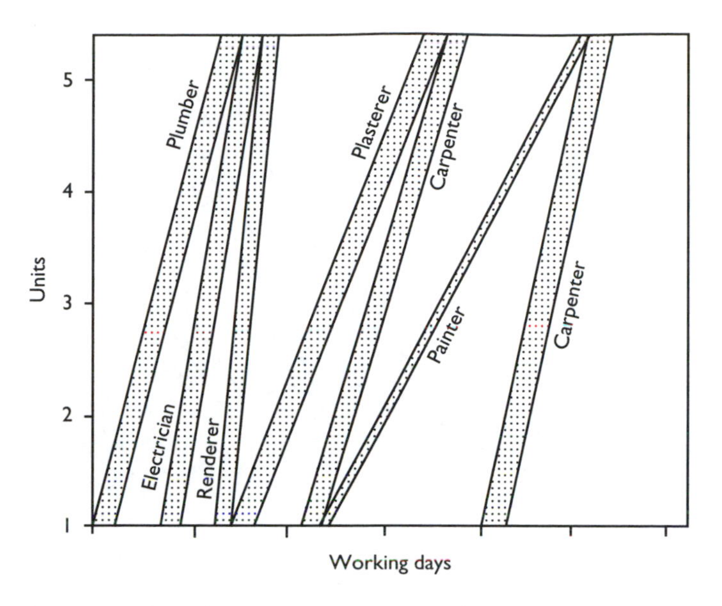

Figure 3.12 Example project with sequential/repetitive work. (Activity names have been abbreviated because of space limitations – work items are implied.)

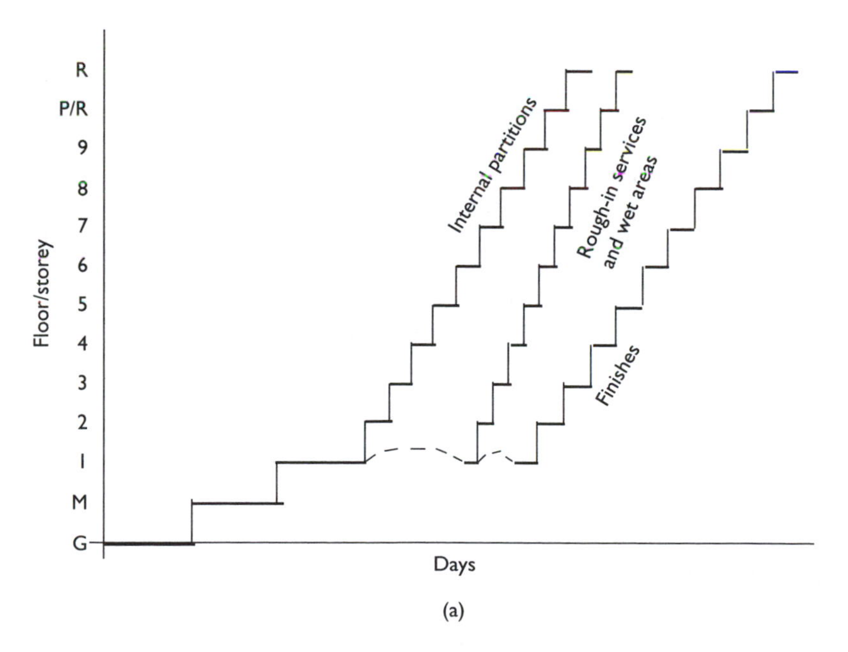

Figure 3.13 Example construction program for internal work for a building. (a) Bar chart; (b) Cumulative version. (Activity names have been abbreviated because of space limitations – work items are implied.)

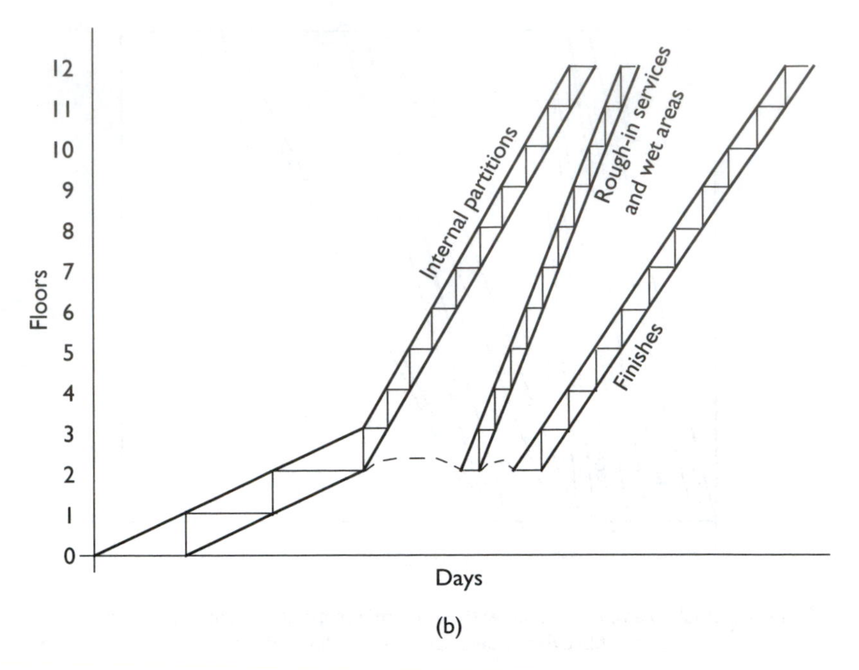

Figure 3.13 (Continued).

(Refer 'resource balancing'.) Associated with every cumulative production plot, a network can be drawn showing the work involved; see 'resource balancing'.

When 'integrated', a bar chart becomes the same as a time-based cumulative production plot. Cumulative production plots, bar charts and multiple activity charts contain essentially the same information. It is the additional *vertical scale* that makes the plot attractive, compared to a bar chart. The computational path to the cumulative production plot should be no different to that to a conventional bar chart, namely via a network.

In cumulative production plot terminology, a *resource schedule* is one determined by resource numbers, while a *parallel schedule* is one determined by production.

Cumulative production plots also find use in reporting for replanning purposes.

Decompression

Decompression refers to the project duration being expanded. The project decompression problem is a subproblem of the total planning problem, where emphasis is placed on the different level controls that can lead to activity and project lengthening, and all other controls are held fixed. As with the case of applying additional resource numbers or using higher resource production rates to decrease (compress) an activity's duration, so applying less resource numbers and decreasing the resource production rates increases (expands) an activity's duration. An expanded project duration may be possible if a later completion time (extension) is available or allowed.

Delay

A delay is a time period affecting the undertaking of or impeding the progress of a project. The causes of delays are many, but only refer to those where uncertainty is involved.

A delay allowance, delay contingency or 'time' reserve may be incorporated into a project's schedule where this is possible. In Figure 3.14, q days have been allowed over the total project. The diagonal line represents a uniform rate of usage of the delay allowance over the total project.

Figure 3.14 is a project level representation.

Deliverable

The term 'deliverable' is loosely used and abused. Its consensus meaning is that of end-product plus state final (terminal) conditions for the project, or equivalent for a project milestone. However, because of the term's abuse, it is best avoided.

In terms of work breakdown, an end result has no meaning, and so deliverables should not appear in a work breakdown structure, in spite of how useful a deliverable might be to some people for project management purposes.

Refer Carmichael (2004).

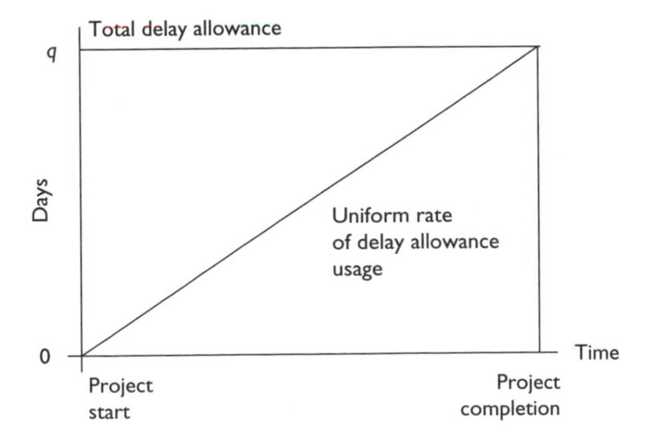

Figure 3.14 Delay allowance; cumulative delay plot.

Disturbance

The corruption that prevents any planned for project output being obtained exactly is represented by disturbance. Disturbance leads to the need for continual monitoring and reporting of project progress, and possibly revised controls.

Earned value

At any time during a project, earned value is the amount that should have been spent on the production done.

The earned value of production performed is found by multiplying the estimated per cent completion for the production by the planned cost for that production.

$$\text{Earned value} = (\text{Estimated})\ \text{Per cent complete} \times \text{Planned cost}$$

Earned value is reporting at the project level.

Element

An element of an activity represents a portion of an activity equal to the smallest chosen time unit for the project. In many projects, the smallest time unit is a day, but other time units are possible. For an activity with a duration of d, then d activity elements result.

End-product

A project will commonly come about because of an identified need or want for some product, facility, asset, service and so on. This end-product is achieved through a project.

This distinction sometimes causes people confusion and many people are not aware of the distinction, or of the need to make a distinction. For example, people sometimes refer to a building as a project. It is not. The processes that go together to materialise the building are the project. The building is the end-product. (It is acknowledged that the definition of a project is sufficiently flexible to include the operation and maintenance phase of a product within what is called the project. However, this is not the issue here.).

Refer Carmichael (2004).

Error control

Closed loop control may be referred to as error control if the control is selected based on the difference between desired system behaviour and actual system behaviour.

Estimates

One distinction between project costs is that between direct and indirect costs. *Direct costs* are costs that are directly proportional to the work quantity. The cost of people, equipment and materials needed to carry out an activity is an example. The costs discussed in network analysis are generally direct costs. *Indirect costs* are one-off and time varying, and accrue whether or not any work is done. Examples include the salaries of project administrative staff, hire of accommodation and insurance. The total project cost is the sum of all the activity direct costs and indirect costs. If developing a tender price, a contingency might be added to account for uncertainties. (Also included would be head office overheads and profit [Figure 3.15].)

Estimates for durations and costs to carry out activities are usually based on past experience (extracted from historical records or 'time' studies) and trends, or could be developed from first principles. The usefulness of estimates depends on the skills of the estimator and the familiarity this person has with the work involved in the project.

Several *degrees of estimate accuracy* may be recognised. Generally, the accuracy of estimates improves as the project stages progress. Early on, estimates may be quite coarse. At implementation, reasonably accurate estimates are required.

A common (but not the only) differentiation of degrees of accuracy in estimates is:

- *Approximate (within 20–25% accuracy).* Usually used to assess the magnitude of the project and based on duration or cost indices.
- *Preliminary (within 10–20% accuracy).* These are compiled from overall unit rates for each piece of work in the project, and are usually used for a decision whether to proceed with the project or not.
- *Detailed (within 5–10% accuracy).* These are used for tendering or replanning purposes, and they require either accurate unit rate methods or a bottom-up approach.

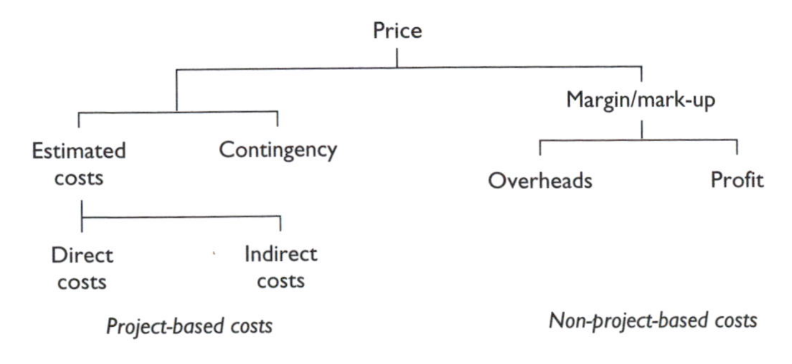

Figure 3.15 Tender price composition.

Estimates depend on resource assumptions, local and production factors including the availability of people and special skills, availability of materials, weather and hazard facts, and availability of equipment/plant.

According to Murphy's Law (extended): *Everything takes longer than you think*. And *Nothing is as easy as it looks*.

Event

An event is the start or end of an activity; an occurrence at a specific time. Special or important events may be given names, such as milestone event. Events correspond to the nodes in activity-on-link diagrams.

Some computer packages may (for computational reasons) represent an event as an activity of zero duration, but this can lead to misunderstanding; an event is not an activity.

In terms of work breakdown, an event has no meaning, and so events should not appear in a work breakdown structure, in spite of how useful an event is for project management purposes.

Factual network

Also called as-executed (as-built, as-constructed, as-implemented) programs. A factual network is a network (best done on a timescale) of as-executed activities indicating changes to the original program, including delays and variations, and agreed changes, cross referenced to files on this information. It is a record of facts. The factual network becomes a permanent record of a project's progress. A factual network (as opposed to the original network) may not lend itself to analysis via conventional methods. The factual network also finds a use in attempting to resolve disputes involving delays, schedule extensions, liquidated damages and related claims. It is similar in concept to an as-built drawing, in that it shows the actual durations to perform the various work items.

Fast-track project

A project is said to be fast-tracked if its phases overlap (Figure 3.16). A phase is started before the previous phase is complete. Project phases are run in parallel where they may normally be run sequentially.

At the activity level, a similar situation can occur when start-to-start, finish-to-finish or start-to-finish relationships are specified on the activity network, in place of more usual finish-to-start relationships. This can have the effect of completing the work sooner, which is the intent of fast-tracking.

The primary benefit of fast-tracking is a reduced schedule, and earlier completion time for the project, that is earlier delivery of the end-product. The actual project cost could be expected to increase, in return for an earlier

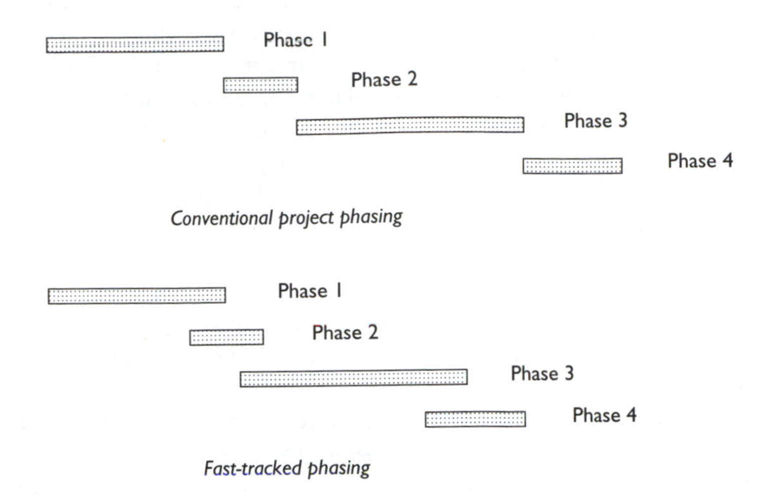

Figure 3.16 Bar chart showing comparison of conventional and fast-tracked approaches.

return on money invested, for example earlier collection of rental, earlier sale and decreased cost of borrowed money.

The downside with fast-tracking is the associated peculiar managerial problems.

Refer Carmichael (2004).

Financial planning

Financial planning refers to all that involved in deciding the usage of money on a project. The term 'budgeting' may also be used to mean financial planning, and the term 'budget' to mean financial plan.

The financial planning problem is a subproblem of the total planning problem, where emphasis is placed on the different level controls that can influence money usage (including recoupment), and all other controls are held fixed. Resource usage is converted to a unit of money.

Budgeting for projects, some might argue, is a more difficult task than budgeting for a business. Projects are unique and have to be developed from first principles each time, compared with a business where next year's operations could be expected to bear some similarity with this year's operations. Though budgeting in a business via work packages (and later monitoring against these work packages) is little different to that involved in budgeting for projects.

Budgets tie expenditure to activities. If this is related to an accounting system, then as work is done, this can be charged against individual account numbers.

All the different participants in a project – owner, contractor, subcontractor,... – have money being expended and various forms of income. Financial planning takes into account expenditure, income and cash flow. For example, where a contractor (and similarly for a consultant) is being reimbursed by the project owner for work done, a *cumulative income* plot can be superimposed on the cumulative expenditure plot to give Figure 3.17. Refer 'S curve'. Incomes and expenditures might be plotted based on dates of anticipated transfer of funds and billings, when invoiced, when committed, or when paid, depending on the user.

The comments below apply for both contractors and consultants.

Commonly project work is done by contractors, but not paid for by the owner until progress claims or similar are submitted. The contractor funds the project.

The vertical difference between the cumulative outlays of the contractor and the cumulative reimbursements represents the amount of money the contractor has to find on any day to finance the project. It may not be until the end of the project that the contractor shows a profit. Many projects have a negative cash flow until the very end when final payment, including retention funds, is received.

There may be large differences between cash flow patterns on different projects. A positive cash flow is attractive to a contractor since it eliminates borrowing or tying up its own funds (which then become available for investment elsewhere). Negative cash flow draws on the contractor's

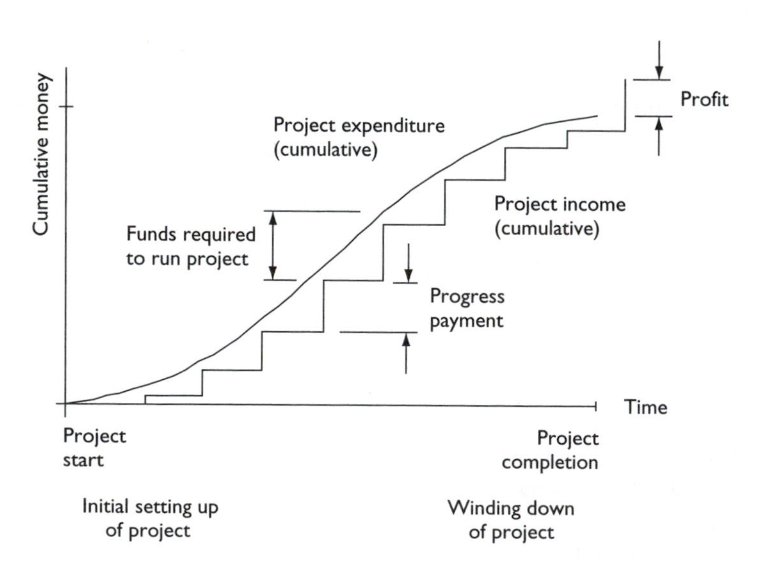

Figure 3.17 Example cumulative expenditure and cumulative income diagrams (as-planned).

working capital, or borrowing is necessary. For a project to proceed, funds availability has to exceed funds requirements.

As shown in Figure 3.17, the contractor is funding the project except for right at the end of the project. The expenditure exceeds the income, implying a negative contractor cash flow. Contractors might attempt to improve their cash flows by moving the cumulative expenditure and cumulative income diagrams closer together. The ideal from the contractor's viewpoint is to have the cumulative income diagram above the cumulative expenditure diagram, implying a positive cash flow for the contractor; the owner is funding the project.

The ways that a contractor might improve its cash flow include:

- Receiving a down payment, up-front payment, advance payment, mobilisation payment, part payment on acceptance of tender, or similar terminology.
- Receiving payment of a progress claim in a timely fashion.
- Delaying payments to creditors, consultants, subcontractors and suppliers to 30, 60, 90 or more days (a lag between when an invoice is received and when payment is made).
- Unbalancing its bid through front end loading. That is, items that occur early in a project are bid and reimbursed at marked-up prices/rates.
- Starting activities closer to their LST rather than their EST. Planned expenditure will commonly lie somewhere between these two extremes.

The inflow of funds is also affected by profit gained on the project, and reworked into the project. Progress payments may be invested and attract interest.

Awareness of all these issues allows the contractor to put together a considered bid, or be selective in its bidding on projects.

From the owner's viewpoint, there may be curiosity in the contractor's cash flow, but possibly more curiosity in the owner's payment schedule which is the contractor's income plot. Owners may try to defer payment to a contractor for as long as it does not hinder the contractor's performance.

Before awarding the work to a contractor, the owner may need to have an idea as to what its total costs for the project will be, and when it will be required to make progress payments (and hence by when money should be obtained).

Feasibility, from an owner's perspective, would generally be established based on a master plan level of generality.

The cost of financing

On large projects, the cost of financing can be a major consideration. Interest on borrowed money can represent a significant contribution to the project cost. Therefore, expenditure tends to be postponed for as long as possible, subject to not having a deleterious effect on a project's progress.

On the other hand, *escalation* costs are hard to predict. (Commonly, an escalation allowance is calculated for the midpoint of the time span over which the project work is expected to be performed. Escalation is assigned to each work package relative to the duration of each package.) The earlier a project is completed, the cheaper will be the associated end-product, and end-product income can start to be received. To guard against escalation, components may be purchased early, but this introduces *inventory* costs, which include deterioration, damage and even theft while in storage. Cash flow calculations accordingly should include interest, escalation and inventory costs.

Consideration of all these issues will lead to a minimum cost solution.

Float

Also called slack. The term 'float' is used in this book. Float is a free time period. Float is the time period available to delay the start or finish of an activity without affecting the project duration. Conventionally, non-critical activities have float; critical activities have no float.

Several types of float may be distinguished. The most popular are *total float* (TF), *free float* (FF) and *interfering float* (IF). They are related through,

$$\mathrm{TF} = \mathrm{FF} + \mathrm{IF}$$

The definitions of all floats are similar. All are time periods available to delay the start or finish of an activity without affecting the project duration. Total float is the total time period available. It may be thought of as being composed of free float and/or interfering float. Free float is a time period available, but if used will not affect the start times of following activities. Interfering float is a time period available, but if used will affect (interfere with) the start times of following activities.

Whether float is classified as free float or interfering float depends on where the activity occurs in the network. For example, consider the part activity-on-node network of Figure 3.18. The critical path runs through the bottom path and has a duration of 12 days. The top path has a combined duration of $3+2+4=9$ days, giving a total float of $12-9=3$ days. This 3 days appears as interfering float in activities A and B, but as free float in activity C. It is the same float manifesting itself in different ways, and may be used in activities A, B and C in any combination, but only up to a combined total of 3 days. It is called interfering float in activities A and B, because if used in A and/or B, it affects the start time of C. It is called free float in C, because if used in C, it does not affect the start time of any other activity.

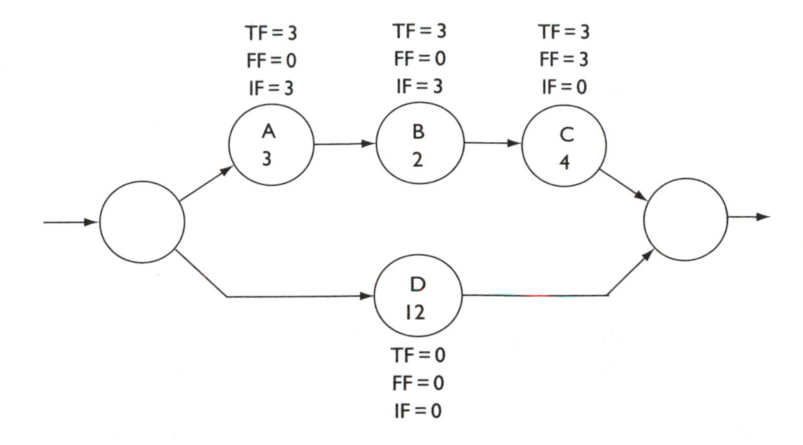

Figure 3.18 Part-network illustrating float type. Activities are indicated by the letters A, B, C and D, and their durations are given within the circles.

Other floats are mentioned by various writers, for example independent float and scheduled float.

Where the float is negative, this implies that the required project completion date is sooner than the project completion date obtained through a critical path analysis. Activity durations will have to be reduced by this float amount for the required completion date and calculated completion date to be the same.

Gantt chart

Refer 'bar chart'.

Human resource planning

Human resource planning refers to all that involved in deciding the usage of people on a project.

The human resource planning problem is a subproblem of the total planning problem, where emphasis is placed on the different level controls that can influence the usage of people resources, and all other controls are held fixed.

Lead time period

Also called lead time and lag. It is denoted LT in this book. Refer 'overlapping relationships'. Lead time period refers to the duration between starting/finishing a preceding activity and starting/finishing a following activity.

Lead time period may be expressed as an absolute value, in days, for example, or as a percentage of the duration of an activity. When expressed as a percentage, then reducing the activity duration will reduce the associated lead time period; this situation, of course, does not happen where lead time periods are expressed as absolute values.

Note that some writers make a distinction between 'lead' and 'lag'. That distinction is not followed in this book.

Learning

As an individual, team or organisation gains experience in carrying out its allotted work, the effectiveness of the individual etc. increases. There is a saying 'practice makes perfect'. Typically the time period, for example, to make or do a first unit is higher than the time period to make or do a second unit and this time period continues to decline for further units though to a slightly lesser extent.

Generally, as a task is repeated, performance improves with a tapering off after carrying out the task many times. Knowledge of such possible improvements in performance can prompt a redesign of the work method; the resultant time period savings may be incorporated into the plan or kept as a reserve.

Figure 3.19 shows this schematically and is called a learning curve. The curve is applicable for individuals, teams and organisations and has been substantiated in many field observations. Such information has direct relevance to planning any repetitive process, or converting processes into repetitive ones.

In bar chart form, the learning effect might be represented as in Figure 3.20.

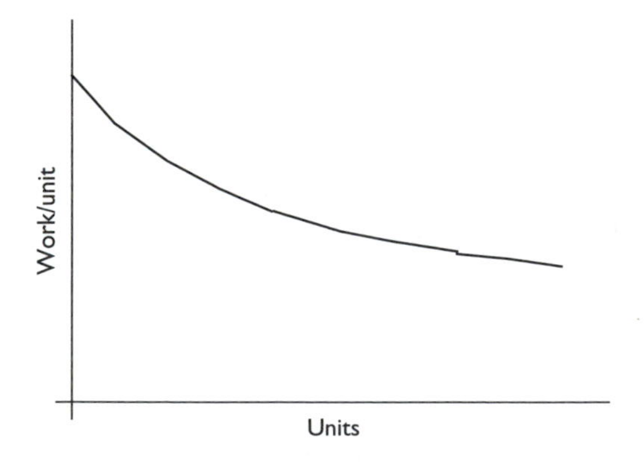

Figure 3.19 Learning curve.

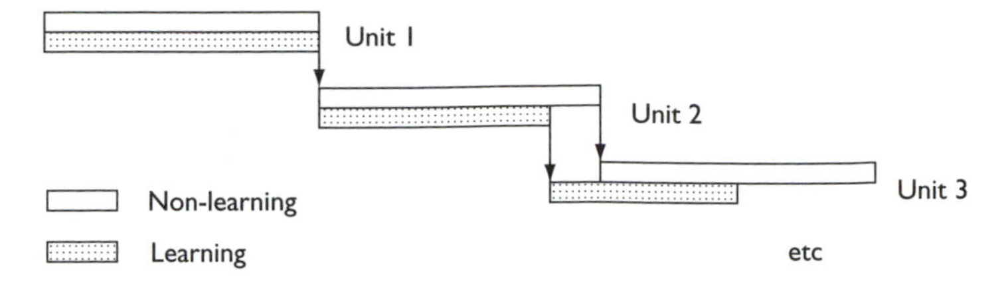

Figure 3.20 Bar chart representation of learning effect. The bars refer to the work involved with each unit.

The learning curve is commonly described by an equation of the form,

$$T_n = T_1 n^b$$

where

T_n average work/unit after n units
T_1 work/unit for the first unit
b index of learning ($b < 0$)

When plotted on logarithmic coordinates, this exponential curve becomes a straight line (Figure 3.21).

Different values of b, the index of learning, represent different values of the rate of learning as shown in Figure 3.22.

Conventionally, however, the rate of learning is specified as a percentage such as 70%, 80%, 90%,... For example, an 80% curve means that the work involved with the nth unit is 80% of that required for the

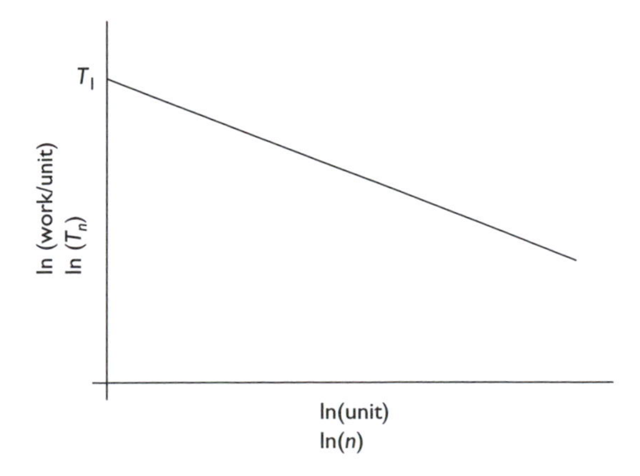

Figure 3.21 Learning curve on logarithmic scales.

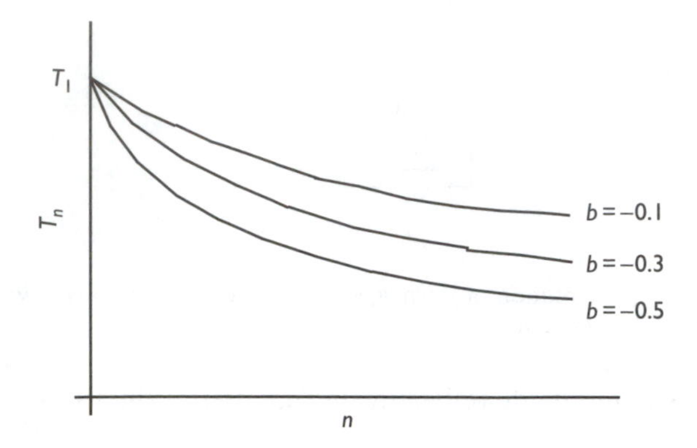

Figure 3.22 Family of learning curves.

($n-1$)th unit. 70%, 80% and 90% curves correspond to b values of about -0.51, -0.32 and -0.19 respectively.

Both T_1 and b require estimating. Typically, these are adapted from similar situations or historical records.

Level

A project may be decomposed to subprojects, to sub-subprojects, . . . , to activities, elements and constituents. Levels refer to the strata in such a hierarchical decomposition.

Line of balance technique

The line of balance technique applies to resourcing on projects that typically have sequential/repetitive work content (for example, the work of trades in constructing a high-rise building). It is a process-oriented approach which has its origins in the manufacturing industry. Refer 'resource balancing', and 'cumulative production plot'.

Milestone

A milestone is an important event occurring in a project. A project may have a number of milestones. Milestones are shown on a network or bar chart in order that everyone is aware when particular important events occur, or something should be accomplished by.

As a computational and representational device, some computer packages may define a milestone by setting a bogus activity's duration to zero, but nevertheless a milestone is an event not an activity.

Monitoring

Monitoring involves measuring or observing the output or performance of a project.

Monte Carlo simulation

Monte Carlo simulation, as used in planning, is a numerical probabilistic method of network analysis. However, note that Monte Carlo simulation is a technique for analysing nearly anything that contains probabilities.

The currently popular way to realistically incorporate variability in activity durations (typically represented by probability distributions) for network analysis is to use Monte Carlo simulation.

Monte Carlo simulation essentially converts a difficult probabilistic problem into many simpler deterministic problems. The steps involved for network analysis are:

- Activity durations are sampled (via the generation of uniform random numbers).
- A conventional network analysis (incorporating overlapping relationships if applicable) is carried out.
- The above two stages are repeated many times.
- Relevant statistics on important items, such as project completion times, are collected.

The first step is one of data generation, the second step is analysis, while the fourth step is bookkeeping.

Monte Carlo simulation is suited ideally to computer use, where large amounts of data can be handled with ease. It is not a technique for use by hand or where a closed-form solution is required.

Some commercially available Monte Carlo simulation packages sit as overlays on critical path method (CPM) packages.

Multiple activity chart

Also called a multi-activity chart, it is a chart on which the activities of more than one resource (people or equipment) are recorded on a common timescale to show their interrelationship.

A multiple activity chart is drawn in a similar sense to bar charts with a timescale and shading length in proportion to the duration of whatever is being plotted. Separate bars are drawn for each resource. The bars may be drawn vertically (top to bottom time sense) or horizontally (left to right time sense).

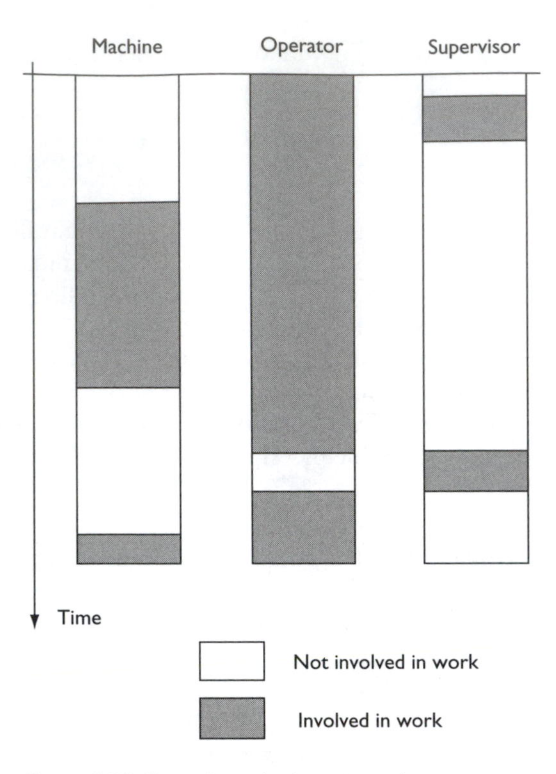

Figure 3.23 Example multiple activity chart.

The chart shows very readily periods of idleness and periods when the resource types are utilised.

Figure 3.23 shows an example of multiple activity chart. Times for such charts are typically based on field records.

A multiple activity chart is essentially a different form of bar chart concentrating on what particular resource types are doing, usually without connections, but there is no reason why connections cannot be included. Bar charts tend to be developed beforehand, while multiple activity charts tend to show what has happened; apart from that, they are no different.

Cumulative production plots, bar charts and multiple activity charts contain essentially the same information.

Murphy's Law

The origin of Murphy's Law, the original principle '*If anything can go wrong it will*' is unclear. Some of the published attribution is not convincing. Possibly the naming came about because Murphy is a distinctly Irish name, and the Irish are known for their quirky humour. Many of the following examples are from Murphy's Law, A. Bloch, Methuen, London, 1986.

Nagler's comment on the origin of Murphy's Law
Murphy's Law was not propounded by Murphy, but by another man of the same name.

The reliability principle
The difference between the Laws of Nature and Murphy's Law is that with the Laws of Nature you can count on things screwing up the same way every time.

O'Toole's commentary on Murphy's Law
Murphy was an optimist.

Goldberg's commentary
O'Toole was an optimist.

There are many extensions and corollaries of the basic principle behind Murphy's Law, and again the authorship of these is unclear. All give a humorous view of laws guiding life.

Murphy has something to say on planning related matters. Some examples are:

No matter what goes wrong, it will probably look right.

Nothing is as easy as it looks.

Everything takes longer than you think.

Before you can do something, you have to do something else.

The nearer to completing the job, the greater the alterations required.

Now they tell us Law
Information calling for a change in plans will arrive after the plans are complete.

Sodd's First Law
When a person attempts a task, he or she will be thwarted in that task by the unconscious intervention of some other presence (animate or inanimate). Nevertheless, some tasks are completed, since the intervening presence is itself attempting a task and is, of course, subject to interference.

Pudder's Law
Anything that begins well, ends badly.
Anything that begins badly, ends worse.

Wynne's Law
Negative slack tends to increase.

3rd Corollary to Law of Applied Confusion
After adding two weeks to the schedule for unexpected delays, add two more for the unexpected, unexpected delays.

Laws of Computerdom according to Golub

1. *Fuzzy project [goals] are used to avoid the embarrassment of estimating the corresponding costs.*
2. *The carelessly planned project takes three times longer to complete than expected; a carefully planned project takes only twice as long.*
3. *The effort required to correct course increases geometrically with time.*
4. *Project teams detest weekly progress reporting because it so vividly manifests their lack of progress.*

Patton's Law
A good plan today is better than a perfect plan tomorrow.

Frothingham's Fallacy
Time is money.

Westheimer's Rule
To estimate the time it takes to do a task: estimate the time you think it should take, multiply by 2, and change the unit of measure to the next highest unit. Thus we allocate 2 days for a one-hour task.

Ninety–Ninety Rule of Project Schedules
The first ninety percent of the task takes ninety percent of the time, and the last ten percent takes the other ninety percent.

Cheop's Law
Nothing ever gets built on schedule or within budget.

The Einstein Extension of Parkinson's Law
A work project expands to fill the space available.
Corollary. No matter how large the work space, if two projects must be done at the same time they will require the use of the same part of the work space.

Workshop Principle (number 4)
The more carefully you plan a project, the more confusion there is when something goes wrong.

McDonald's Corollary to Murphy's Law
In any given set of circumstances, the proper course of action is determined by subsequent events.

Seay's Law
Nothing ever comes out as planned.

Sweeney's Law
The length of a progress report is inversely proportional to the amount of progress.

First Law of Corporate Planning
Anything that can be changed will be changed until there is no time left to change anything.

Kent Family Law
Never change your plans because of the weather.

The following Murphy's Calendar is useful for planners.

FRI	*MON*	*TUES*	*WED*	*THURS*	*MON*	*SAME DAY*
7	6	5	4	3	2	1
14	13	12	11	10	9	8
21	20	19	18	17	16	15
28	27	26	25	24	23	22
35	34	33	32	31	30	29

This calendar has certain advantages:

- Every job is a rush job. Everyone wants things done yesterday. With this calendar an order given on the 7th can be carried out on the 3rd.
- Everybody wants things done early – on Mondays for preference. So there are two Mondays in each week.
- There are several extra days at the end of the month for those end of the month rushes.
- There are no bothersome non-productive Saturdays and Sundays.
- There is a new day each week – 'same day'. On this day, 'while you wait' and 'same day' jobs may be handled without interruption to other promises. Everyone will be happy and ulcer free.
- Acceptance of this calendar is made easier by pointing out that this system has been in unofficial use in many companies for some years.

Network

A network is a diagram showing the logical sequence of undertaking activities. In other words, it shows the precedence relationships between activities, the work order interrelationship of the activities, or the logic of how the project is to be undertaken. In drawing networks, care has to be exercised that false logic is not introduced.

Networks are the basis of project planning and, as part of an iterative analysis approach to planning, represent a rational intermediary step to project programs, resource and expenditure plots and S curves.

In general, a network is composed of *links* and *nodes*. For networks used in planning, the links are directional and are represented by arrows; the network may accordingly be called a *directed network*. This gives rise to the two fundamental types of networks:

- *Activity-on-link diagrams* (elsewhere called arrow diagrams or activity-on-arrow diagrams) whereby activities are represented by arrows starting and finishing at nodes (also called starting and finishing *events*) (Figures 3.24a and 3.25a).
- *Activity-on-node diagrams* (elsewhere called precedence diagrams or circle diagrams) whereby activities are represented by the network nodes, with arrows linking the activities (Figures 3.24b and 3.25b).

Terminology used in this book is activity-on-link diagram and activity-on-node diagram, and both are collectively referred to as *precedence diagrams*, even though this is not the practice in other publications which generally use conflicting terminology.

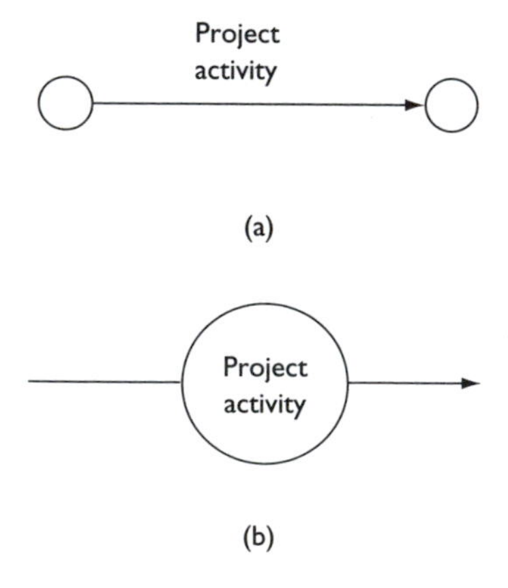

Figure 3.24 Network diagram types. (a) Activity-on-link diagram; (b) Activity-on-node diagram.

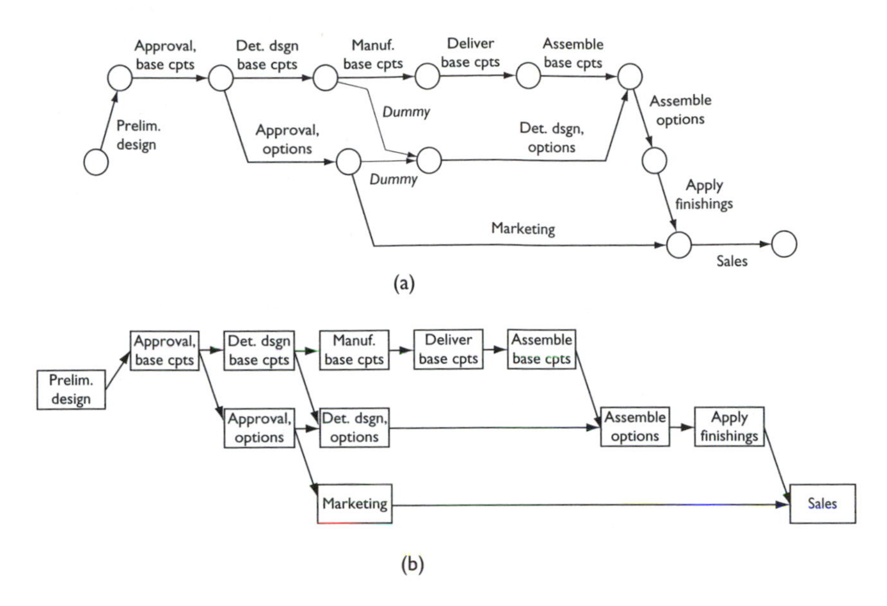

Figure 3.25 Example networks. See the associated bar chart, Figure 3.1, and time-scaled network, Figure 3.55. (a) Activity-on-link diagram; (b) Activity-on-node diagram. (Activity names have been abbreviated because of space limitations – work items are implied.)

The same analysis is used for both network types, and the two types of diagrams give the same answers. It is possible to develop a transformation between the two types of diagrams.

It is a matter of personal preference which network type it used, although the orientation of computer packages accessible to the user is an influence. Clients may also insist on the use of a particular computer package which in turn will often determine the type of network used. Some computer packages allow both types of diagrams. Activity-on-link diagrams were the basis of early critical path analysis work in the 1960s and 1970s. Recent favour by users seems to be towards activity-on-node diagrams.

Some people insist that activity-on-node diagrams are much better than activity-on-link diagrams, yet the same people use connected bar charts to convey a project's program; that is, both activity-on-node and activity-on-link diagrams are being used simultaneously. Many unknowingly use activity-on-link diagrams (to a timescale) when using certain software packages; the activity-on-link diagram might be disguised as a connected bar chart – this is in fact a time-scaled activity-on-link diagram.

In favour of activity-on-link diagrams, bar charts are universally used, and activity-on-link diagrams are the only type that can be represented to a timescale.

In favour of activity-on-node diagrams, they allow easier modification (editing or updating) once initially developed. Adding and taking away links in activity-on-node diagrams tends to be easier than adding and taking away nodes in an activity-on-link diagram. The activity-on-node diagram tends to be easier to understand and follow by all strata of management and permits an easier numbering or coding scheme. An activity can be connected to other activities directly; with an activity-on-link diagram there may be a need to introduce a dummy activity in order to not affect the required logic. For overlapping relationships, such relationships apply between activities and these are easier to represent on an activity-on-node diagram.

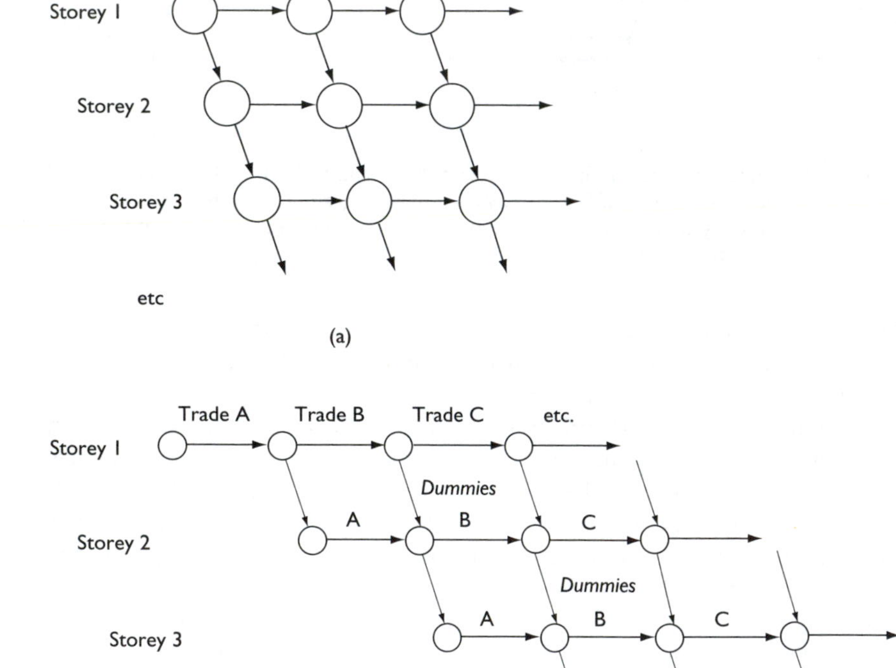

Figure 3.26 Part activity-on-node and activity-on-link diagrams applying to the construction of a multistorey building. (a) Activity-on-node diagram; (b) Activity-on-link diagram; the vertical arrows between storeys are dummy activities. (Activity names have been abbreviated because of space limitations – work items are implied.)

There is a special case where activity-on-node diagrams are preferred to activity-on-link diagrams, and that is where there are the same sequential activities repeating. See for example, Figure 3.26, which relates to the construction of a multistorey building; each trade in sequence works its way up the building. The activity-on-node diagram turns out to be a more economical representation compared to the activity-on-link diagram, which needs the incorporation of many dummy activities to maintain the logic.

For other 'linear' type projects, for example relating to roads, pipelines, railway lines and like structures involving sequential/repetitive activities, activity-on-node diagrams may also be preferred.

In drawing a network:

- The diagram is not drawn to any scale.
- The diagram has a left to right sense, where the project start is at the left, and project end is at the right. Activities to the left of the diagram could be expected to occur before activities to the right of the diagram.
- Arrows do not have to be straight. The length and direction of the arrows is not important, but generally have a left to right sense. Where arrows cross over, pipeline style drafting can be used to help reduce possible confusion.
- Activities can be located anywhere vertically on the page. It may be found convenient though to locate activities associated with a certain trade or function together, and also to locate the important activities near the (vertical) centre of the page.
- Various drafting styles are possible for nodes, ranging from circles to multicelled rectangles. The size and form of the shapes are not important. Figure 3.9 shows some examples.
- A following activity may depend on a preceding activity(ies) according to any of four overlapping relationships – finish-to-start (F/S), start-to-start (S/S), finish-to-finish (F/F) or start-to-finish (S/F). Multiple relationships may also occur between activities.
- Activities that can be done concurrently are drawn 'in parallel'.
- Activities that are dependent are drawn 'in sequence'.
- Multiple dependency gives rise to multiple links into and/or out of an activity (activity-on-node diagram) or multiple activities into or out of nodes (activity-on-link diagram).
- Dummy 'Project Start' and 'Project Finish' activities can be added for completeness, but are not essential.
- Circular logic or circuits are not possible. Redundant logic can be removed.
- Where numbering (optional) is used for links and/or nodes, the numbering scheme can usually be anything the user chooses, although a computer package may require a specific convention.

Refer Carmichael (1989).

Network analysis

Refer 'critical path method, CPM' (equivalently critical path analysis, CPA), 'PERT' and 'Monte Carlo simulation'.

Objectives

The plural terms 'objectives' and 'constraints' are generally used in this book even though the singular forms may be applicable in some situations; the exception is where the singular is deliberately intended.

The materialisation of a project's end-product can be performed, possibly, in an infinite number of ways. In systems studies terms, the problem is an inverse problem, and hence there is no unique solution.

The criteria by which the preferred materialisation of the end-product is selected are the project objectives. Work method, scope, resource and resource production rate considerations on projects follow. The selection of the preferred materialisation or means to the end-product is the solution to the planning problem.

In a similar idea, there are possibly an infinite number of versions of end-products. The criteria by which the preferred end-product is selected are the end-product objectives. Form, function, finishes etc. of the end-product follow. The selection of the preferred end-product form is a design problem.

That is, on any project there are two types of objectives:

1. End-product objectives
2. Project objectives (Figure 3.27).

Commonly, project objectives say something about project cost, project duration and deviation from specification, but other objectives are possible. And these may apply throughout the project (for example, minimum

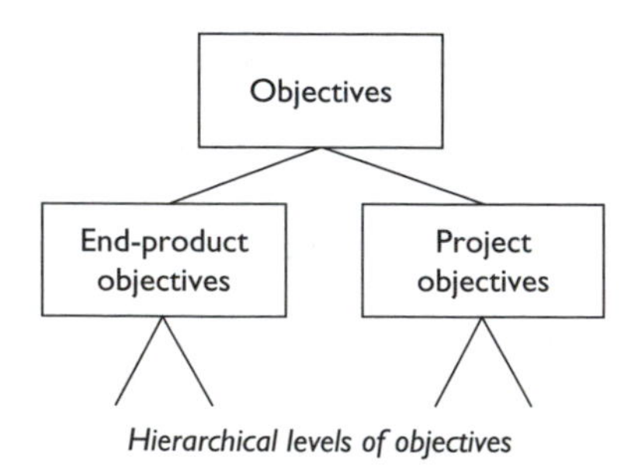

Figure 3.27 End-product and project objectives.

deviation from specification), or at the final (terminal) point of the project (for example, minimum duration which is related to the project completion time). That is, general project objectives will contain a component over the time domain of the project and a component at the final (terminal) point.

Objectives may be expressed at the various project, activity, element and constituent levels.

End-product and project objectives may derive from higher level values within an organisation, for example originating from a corporate plan. Such values may also reflect political, marketing, environmental (natural),... concerns.

End-product objectives and project objectives (and constraints) may relate, for example, to:

- Money (end-product – sales, benefit:cost ratio (BCR), net present value (NPV),...; project – cost, budget,...)
- Duration (end-product – lifetime,...; project – duration, ...)
- Resource usage
- Quality issues
- Community acceptance
- Environmental (natural) effects
- Safety
- Risks
- Public impact
- Extreme event impact (floods, cyclones,...)
- Social impacts
- Geotechnical considerations.

These are expressed in terms of end-product matters or project matters, as the case may be.

Refer Carmichael (2004).

Alternative names for objective. As needed to perform systematic problem solving/systems synthesis, alternative names include:

- (Optimality) criterion
- Performance index
- Performance measure
- Merit function
- Payoff function
- Figure of merit
- Aim

- Goal
- Cost function
- Design index
- Target function.

These generally occur outside the management literature. They are common in the optimisation, optimal design and optimal control literature (Carmichael, 1981).

Open loop control

In open loop control, the control is selected up front, and no follow-up monitoring of system performance and corrective control or action is carried out.

Output

The output describes the external or observable system response, performance or behaviour. For the stripped-back projects considered in this book, output and state are the same, that is there are no observability (in the sense of Kalman) issues. Output contains multiple parts (that is, it is a vector quantity of output variables). Output is a controlled variable.

Overlapping relationships

An elementary development of networks assumes that when one activity finishes another can start (with zero lead time period between the activities). The activities are said to have finish-to-start (F/S) relationships. Generalisations of this are possible to embrace various relationships between the starts and finishes of dependent activities, namely *start-to-start (S/S)*, *finish-to-finish (F/F)*, *start-to-finish (S/F)*, and *finish-to-start (F/S)* relationships with non-zero *lead time periods (LT)* (Figure 3.28). For example, a start-to-start relationship implies the following activity cannot start until some specified time period after the previous activity has started, while a finish-to-finish relationship implies the following activity cannot finish until some specified time period after the previous activity has finished.

Planners seem to have most usage for F/S, S/S, F/F and S/F relationships in that order.

Using S/S, F/F and S/F activity overlapping relationships can be like fast-tracking at the activity level; activities run in parallel instead of sequentially.

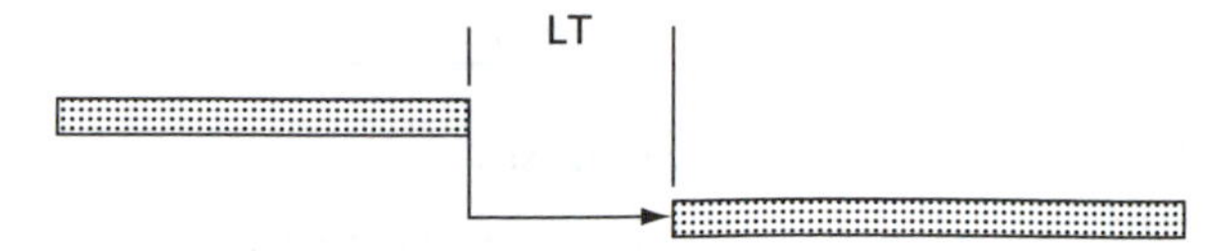

Finish-to-start (F/S) with lead time period

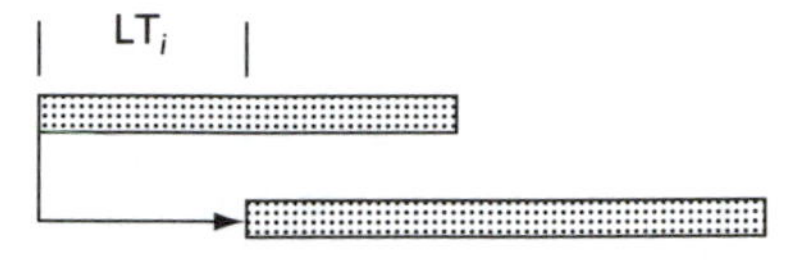

Start-to-start (S/S) with lead time period

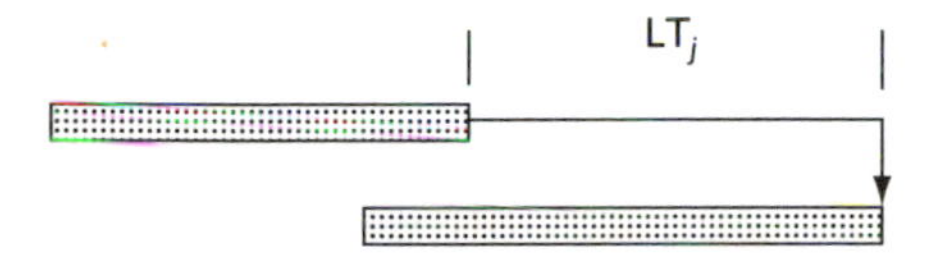

Finish-to-finish (F/F) with lead time period

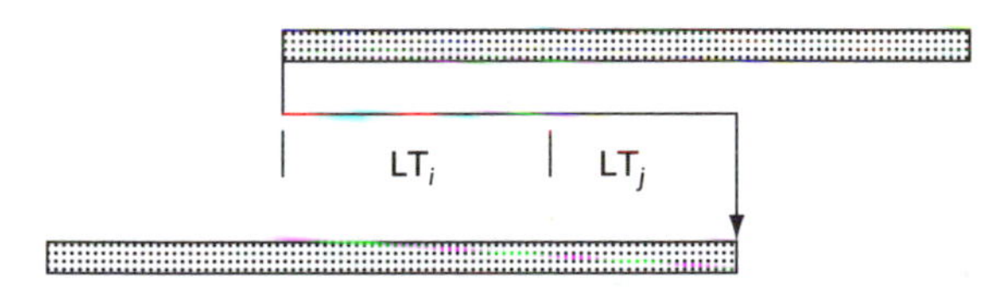

Start-to-finish (S/F) with lead time period

Figure 3.28 Possible overlapping relationships between activities; bar chart representation.

Subactivities

Many people find the overlapping facility helpful, and also economical in terms of representations such as the network diagram, bar chart and so on. However, overlapping relationships can be avoided through the use of subactivities. See for example, Figures 3.29–3.31.

Example

A situation in which start-to-start relationships, for example, find application is illustrated in Figure 3.32, which relates to the construction of a

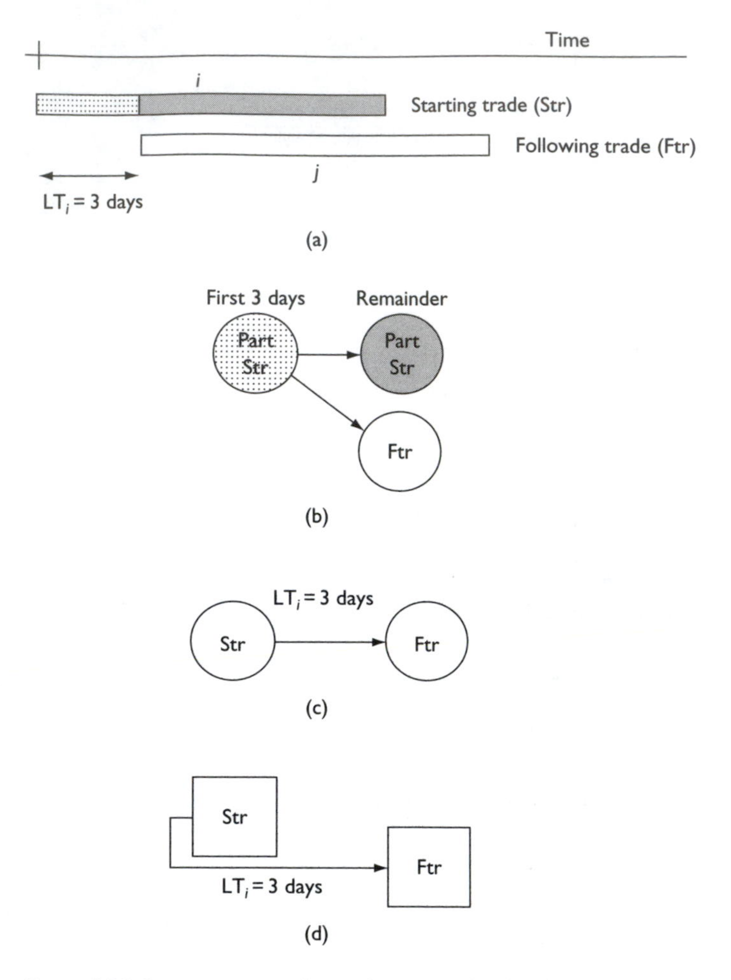

Figure 3.29 Start-to-start relationship example. (a) Bar chart; (b) Without using overlapping ideas; (c) Using overlapping notation; (d) Alternative drafting style. (Activity names have been abbreviated because of space limitations – work items are implied.)

multistorey building; each trade works its way up the building. Start-to-start relationships may apply between trades in order to accelerate the completion of the building; it may not be necessary for one trade to completely finish on a floor/storey before the following trade commences.

For other 'linear' type projects, for example, relating to roads, pipelines, railway lines and like structures, start-to-start relationships may also be preferred.

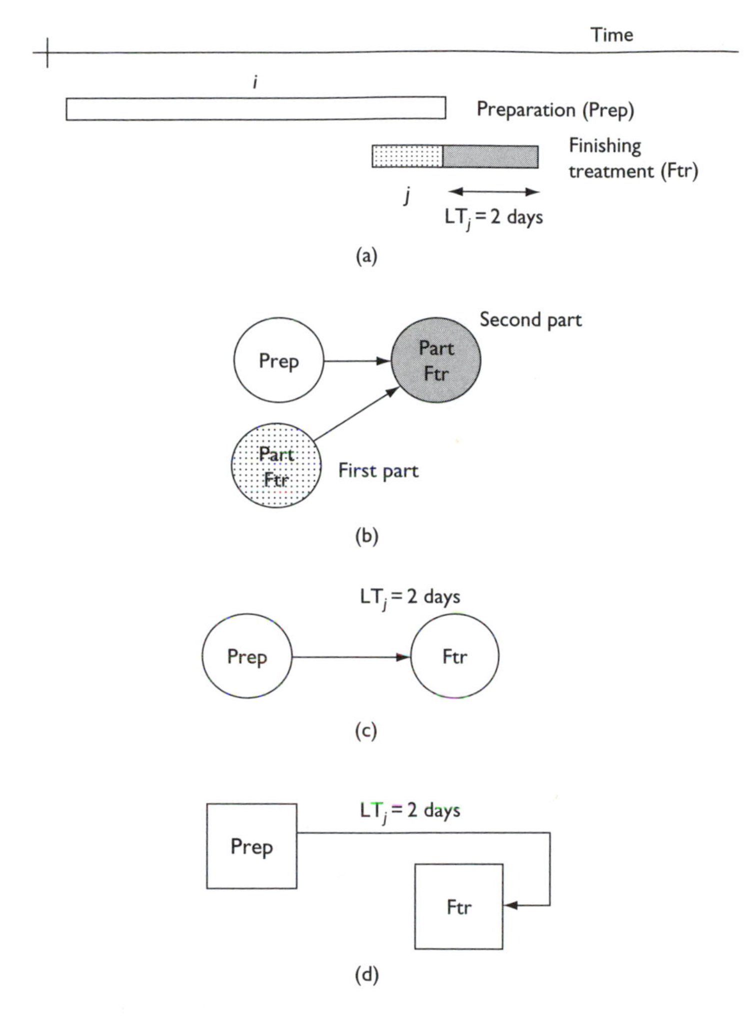

Figure 3.30 Finish-to-finish relationship example. (a) Bar chart; (b) Without overlapping ideas; (c) Using overlapping notation; (d) Alternative drafting style. (Activity names have been abbreviated because of space limitations – work items are implied.)

Example

A situation in which start-to-finish relationships, for example, find application is illustrated in Figure 3.33, which relates to imposing a completion time limit dependent on the start time of the first activity. Figure 3.33a shows the original network. Figure 3.33b introduces a new S/F relationship between the first and last activity. For example, the start of the first activity

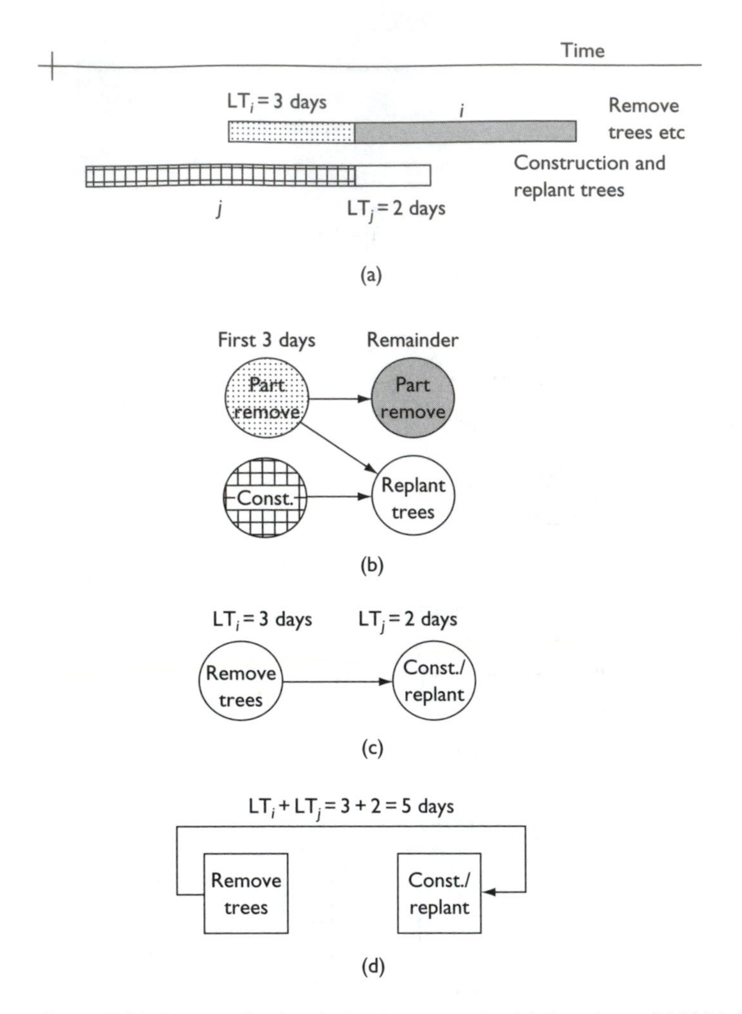

Figure 3.31 Start-to-finish relationship example. (a) Bar chart; (b) Without overlapping ideas; (c) Using overlapping notation; (d) Alternative drafting style. (Activity names have been abbreviated because of space limitations – work items are implied.)

and the end of the last activity may represent milestones constrained by a given time interval between when they occur.

Example

In constructing a building, each floor follows the next. For example, Figure 3.34 shows part of the construction in bar chart form.

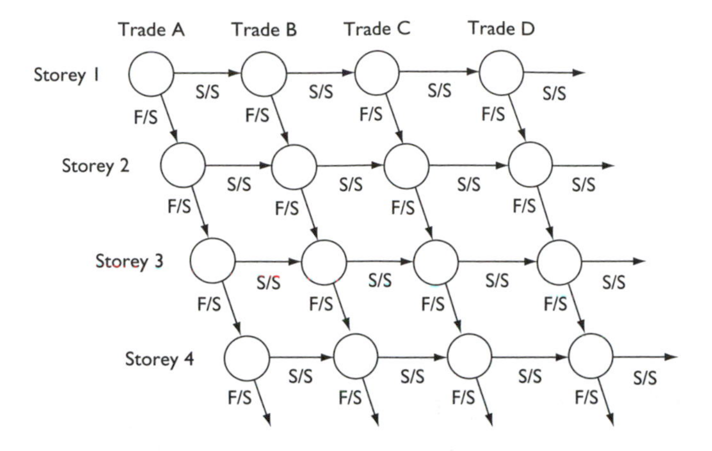

Figure 3.32 Part activity-on-node diagram applying to the construction of a multi-storey building. (Work items implied.)

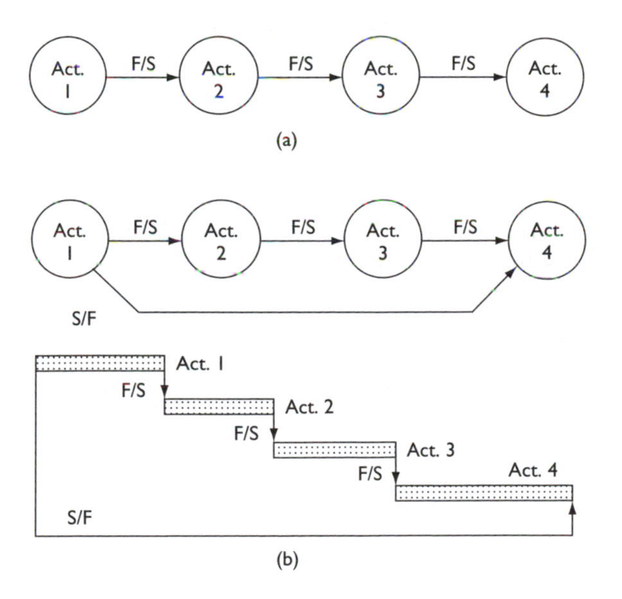

Figure 3.33 Example usage of start-to-finish relationship.

In Figure 3.34a, conventional finish-to-start (F/S) relationships are used.

Figure 3.34b is sometimes termed a *ladder representation*. The second activity has float at its start because the F/F relationship with the first activity dominates. The S/S relationship dominates between the second and third activities.

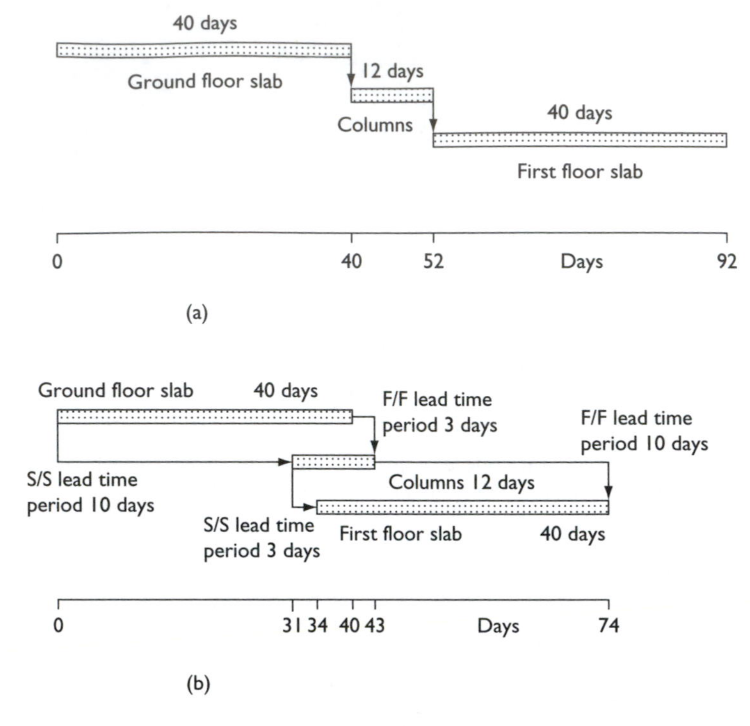

Figure 3.34 Example, building construction. (a) Conventional F/S relationships; (b) Using S/S and F/F relationships. (Work items implied.)

Pareto rule/principle

The Pareto rule/principle is named after Vilfredo Pareto (19th century). It is sometimes called the 80:20 rule/principle. Originally applied to wealth distribution, it has since been generalised: 80% of the outcomes (outputs) are the result of 20% of the influences (inputs). The numbers 80 and 20 are not to be interpreted precisely.

There are many applications in project management. For example, 80% of a project cost estimate is the result of 20% of the component cost items; 80% of a person's output is due to 20% of that person's tasks.

Parkinson's Law

Parkinson's (first) law

Work expands so as to fill the time available for its completion (Parkinson's Law or the Pursuit of Progress, C. Northcote Parkinson, John Murray, London, 1957). The thing to be done swells in perceived importance and complexity in a direct ratio with the time period to be spent in its completion.

Parkinson's (second) law
Expenditures rise to meet income.

Parkinson's law of delay
Delay is the deadliest form of denial.

Performance measure

Performance measures are gauges by which the success or otherwise of a project is measured. A performance measure is in the lay usage sense (but not in an optimal systems sense) of the word 'objective' (mentioned in Carmichael, 2004).

Performance measures tend to be developed later in projects (after the initial planning has been done); objectives are developed very early. A performance measure could be expected to follow planning *baselines* or *targets* (as-planned values) for resource usage (and cost and production). A performance measure may also be based on some industry, competitor or previous project benchmark.

A performance measure, from any of its sources, could be used as the basis for error control thinking.

A performance measure might be termed a key performance indicator (KPI) or similar, but all of these types of terms are used very loosely by most people. The terms are used frequently by project personnel because they sound good and impress, but lack precision.

PERT

PERT (acronym for program evaluation and review technique) is a probabilistic first-order method of network analysis where the duration of each activity is described by a probability distribution derived from an optimistic estimate, a pessimistic estimate and a most likely estimate (Figure 3.35). The associated calculations enable the probabilities of completing events by certain times or keeping to schedule to be calculated. They allow the ranking of activities (an indication of their criticality) according to the probability that the free time period at events is greater than or equal to zero. This information may be used to schedule resources so as to reduce the possibility of delays within a project.

The analysis by PERT is very similar to that involved in the critical path method (CPM – a deterministic method where the durations of activities are estimated to single values) calculations except that PERT carries along one

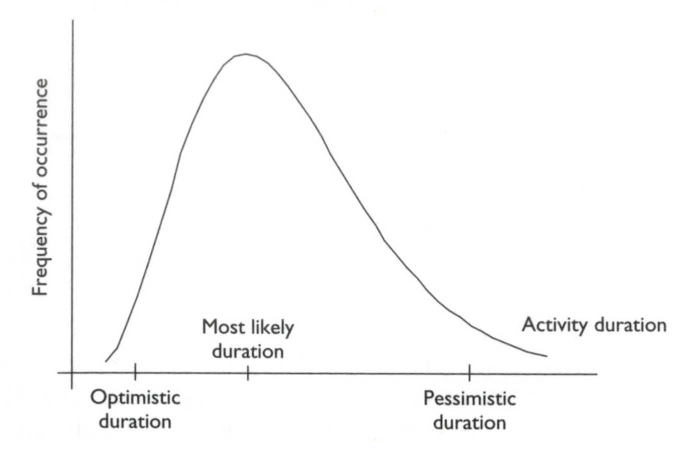

Figure 3.35 Probability density function for activity duration.

additional quantity in the calculations, namely the activity duration variance. That is, PERT calculations work with an activity expected duration and a measure of the scatter or variability in the duration (the variance). (Variance is used here in the probabilistic sense.)

No one appears to be using PERT in practice because such probabilistic approaches are not popular. Probabilistic network analyses appear to be best handled using the technique of Monte Carlo simulation. PERT appears in nearly every textbook on project planning because the technique is useful as a teaching tool to make planners more aware of the assumptions on which deterministic network calculations are based. Planning is an inherently probabilistic problem although in practice it may not be treated as such.

PERT uses an activity-on-link network although events may be given descriptions. It is this characteristic, of distinguishing events along with concentrating the calculations on event times, that can give the misleading impression that PERT uses an activity-on-node diagram as its basis.

Many people and some computer packages use the term 'PERT' but incorrectly – they are in fact meaning CPM. There is much abuse of the term 'PERT'.

CPM might be considered a special (deterministic) case of PERT with a couple of differences. PERT is only developed on activity-on-link networks, and compressing networks in a probabilistic sense raises its own issues.

When the deterministic calculations of the critical path method are carried out, a single project completion time is obtained. To achieve lesser completion times, the project requires compressing. However, PERT gives not only an expected project completion time but also the scatter that may be anticipated in the project completion time. It is thus possible that earlier completion times could be obtained (without the need for compression) if activity durations occur closer to their optimistic durations rather than

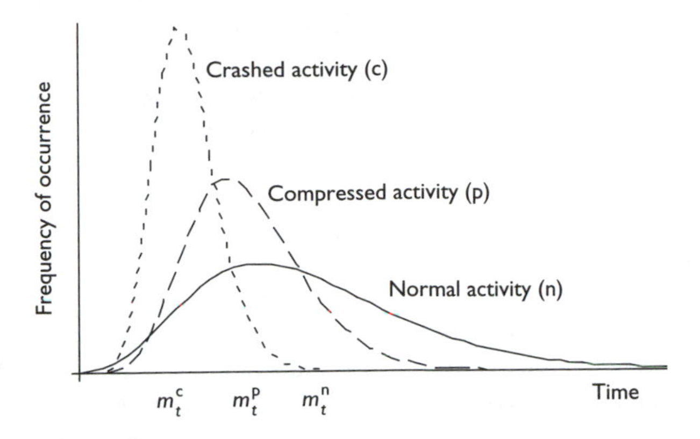

Figure 3.36 Crashing an activity showing mean times m_t.

their most likely durations. A similar argument applies to completion times greater than the expected project completion time.

When network compression is required, it may be carried out similarly to the deterministic case. Corresponding to each expected project completion time there is a variance. Although earlier project completion times may have a finite probability, this probability gets smaller as the project completion time gets earlier. To ensure a higher probability of achieving an earlier project completion time, some compression must be carried out. The situation in Figure 3.36 indicates the same situation for a single activity but the concept is applicable for both activity durations and project durations.

Cost slopes may be defined in terms of activity expected durations and associated costs. In general, network compression using PERT is best used as a guide as to where to apply attention in the network rather than giving absolute results.

Plan, planning

Planning establishes how and what work will be carried out, in what order and when and with what resources (type, and number or quantity, additionally expressed in a money unit). To plan (verb) is the act of choosing the controls (method, resources and resource production rates) throughout the project duration. A plan (noun) is the outcome of planning.

Precedence

Precedence refers to the logical connection or dependence between activities as indicated by a chosen work method. The dependence may be to the start or finish of a whole or part activity.

Precedence may be established, amongst other ways, by addressing the following questions for each activity, in whole or in part:

Q1. Which activities does this activity follow or logically depend upon?
Q2. Which activities does this activity logically precede?
Q3. Which activities can be done at the same time as this activity?

Answers to these questions establish the precedence relationships between activities, and the logic of how the project is to be put together.

The resulting network diagram is no more than the answers to Q1, Q2 and Q3, and represents the logic of how the project is to be carried out.

Generally, if one (B) activity's progress is dependent on another (A) activity's progress, then the network diagram will have activity B connected from activity A. Refer 'network' and 'overlapping relationships'.

Process charts

Process charts portray a sequence of activities diagrammatically by means of a set of process chart symbols to help a person visualise a process as a means to examining and improving it (Currie, 1959; ILO, 1969).

An *outline process chart* gives an overall picture by recording in sequence only the main operations and inspections.

A *flow process chart* sets out the sequence of the flow of a product or a procedure by recording all activities under review using appropriate process chart symbols. A *resource type flow process chart* records how the resource is used.

Process charts are constructed using standard symbols defined in Figure 3.37. The work is broken down into its component activities which are represented schematically by these symbols. Further breaking down of the work may be possible and is sometimes done.

Symbol	Activity	Predominant result
○	OPERATION	Produces, accomplishes, furthers the process
⇨	TRANSPORT	Travels
∇	STORAGE	Holds, keeps or retains
D	DELAY	Interferes or delays
□	INSPECTION	Verifies quantity and/or quality

Figure 3.37 Symbols for process chart construction (Currie, 1959; ILO, 1969).

An *outline process chart* is useful in design and planning. A method study is carried out before any work begins. The chart gives an overall view of the work and graphically shows the sequence of activities where inspections occur and where extraneous materials, information etc. are introduced. It does not show who carries out the work or where the work is carried out. Accordingly, only the OPERATION and INSPECTION symbols are used.

Figure 3.38 shows an example outline process chart. Activities can be numbered in order, and descriptions are placed adjacent to the symbols. Activity durations can also be placed on the chart. Variants on the theme in Figure 3.38 occur, for example, when work is divided or reprocessed, or alternative means are possible to perform an activity; such variants lead to branching, loops and multiple parallel paths respectively.

A *flow process chart* extends the outline process chart to include TRANSPORT, DELAY and STORAGE activities. The chart may be drawn from the point of view of the worker or the equipment or material used by the worker. It is similar in appearance to the outline process chart. Travel distances can be included to scale or by annotation.

Figures 3.39 shows an example flow process chart. Figure 3.40 shows an alternative flow process chart.

References

Currie, R. M. (1959), *Work Study*, Sir Isaac Pitman and Sons Ltd, London.

ILO (International Labour Office) (1969), *An Introduction to Work Study*, International Labour Office, Geneva.

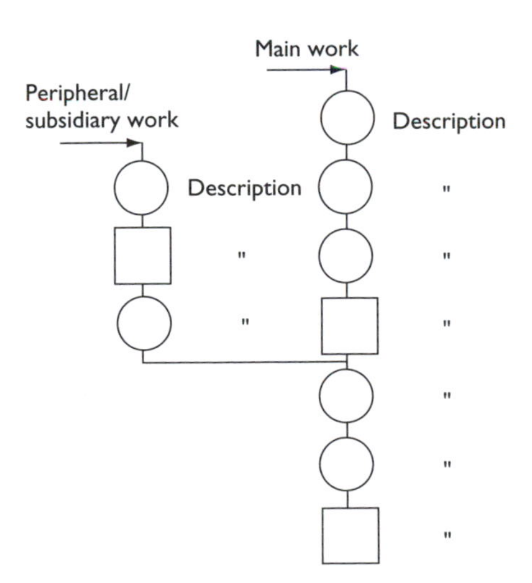

Figure 3.38 Example outline process chart.

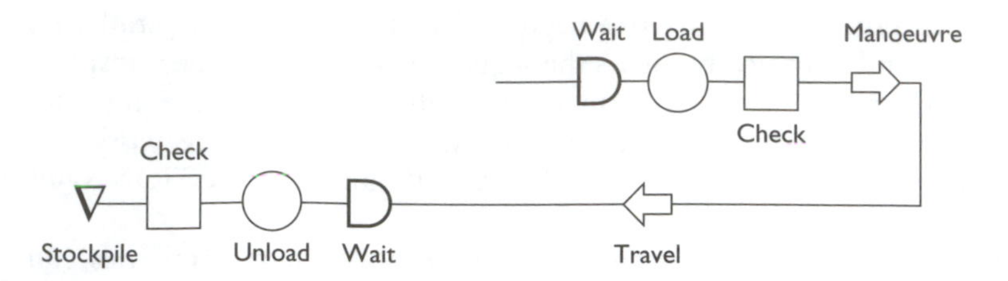

Figure 3.39 Example flow process chart to emphasise distances travelled.

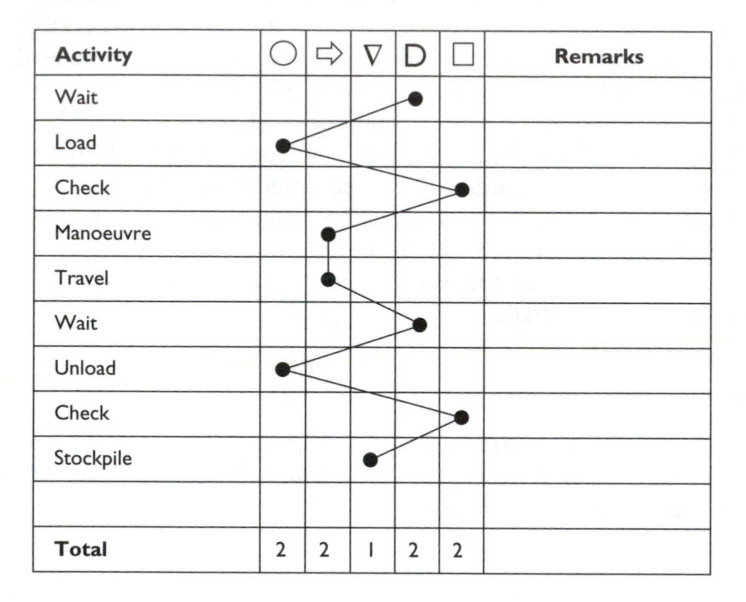

Activity	○	⇨	∇	D	□	Remarks
Wait				●		
Load	●					
Check					●	
Manoeuvre		●				
Travel		●				
Wait				●		
Unload	●					
Check					●	
Stockpile			●			
Total	2	2	1	2	2	

Figure 3.40 Example flow process chart.

Production, production rate

'Production' is used in this book in a general sense to indicate some measure of work done. Production rate is this work done per unit of time. The unit of measurement of production can be various, for example m^2, m or items.

Production rate may refer to a resource's constituent behaviour, or when combined with resource numbers to that achieved in an activity or project.

Program

A program or schedule conveys when work is to be done (and hence it follows, when resources and money are needed). It gives the time relationship of activities, and the sequence of activities needed to achieve the desired project end point. A program contains information at both the activity and project levels. However, enlargements upon this meaning will be found in the literature.

Amongst planners, the term 'program' or 'schedule' may be applied to any or all of the following:

- (Connected) bar chart
- Time-scaled network diagram
- Cumulative production plot
- Activity date/time listings/timetable
- Marked-up drawings.

They typically derive from a network analysis.

A program is presented for the purpose intended. The user has a need for the quick retrieval of relevant information.

The use of the term 'program' here is not to be confused with a computer program; computer programs (commercial software packages) may however be used to generate project programs. The use is also not to be confused with a 'program' referring to a collection of projects; that is a project is a subsystem of a program.

Programs show referenced (as-planned) baselines against which actual progress can be compared. Programs are continually updated, throughout a project, to reflect the latest information available. Sometimes bad practice is seen in that the program becomes an end in itself, and the replanning loop is not closed.

Project

A good definition of a project is hard to find. (Carmichael, 2004). The definition, '*Any undertaking, or set of activities, with starting and ending points, and with defined objectives and constraints, and resource consumption*' is useful, though perhaps unsatisfactory in a number of ways. Finding a satisfying definition of a project is difficult.

Perhaps a more suitable way of thinking about a project would be in terms of attributes that are characteristic of projects, including that they are:

- Unique (one-off, specific discrete undertaking with a unique environment and unique constraints).
- Finite (definable start and end points).

Subprojects. A project is a system and a system can be decomposed to interacting subsystems which are themselves systems. Likewise a project can be decomposed to interacting subprojects (Figure 3.41). Each subproject, which is itself a project, looks after a part of the total project.

Refer Carmichael (2004).

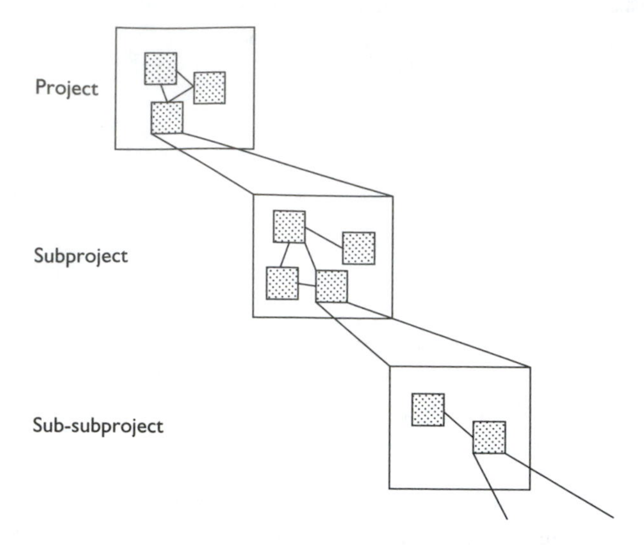

Figure 3.41 Decomposition of a project.

Project management

A good definition of project management is hard to find (Carmichael, 2004).

Many published definitions of project management are partly circular describing management as management. Most definitions are usually in terms of what a project manager does. They often come from studies made of actual projects and observations carried out on the project managers. There are three favoured ways of describing what a project manager does, but other ways do exist:

(i) *A 'classical' management approach.* Management is commonly described in general management texts in terms of:

- 'Planning'
- Organising
- Staffing
- Directing
- 'Controlling'
- Coordinating.

(ii) *A management function approach.* An alternative view of project management is to regard it as a collection or integration of subfunctions:

- Scope management
- Quality management
- 'Time' management

- Cost management
- Risk management
- Contract/procurement management
- Human resources management
- Communication management.

(iii) *Chronological approach*. A project may be seen as going through a number of phases. There is no consensus as to the naming of phases or to the number of phases. In terms of management, many people feel comfortable with a chronological way of describing a project and the activities involved.

Some people find such breakdowns useful. Others find the first two breakdowns so unsystematic as to be of no use. 'Controlling' in (i) above is in the loose lay meaning sense of the word.

Replanning

Replanning gives revised values of the control variables (method, resources and resource production rates) for the remainder of the project. The term 'replanning' is preferred to 'project control' in this book, because lay sloppy usage of the latter has a tendency to mislead, and is inconsistent with the optimum control systems usage preferred in this book.

Reporting

Reporting refers to communicating the results of project monitoring. That is, project output or performance is reported, perhaps in a summarised form. Reporting can take many forms. The form of reporting is chosen to match the intended recipient.

Reserve

Also called *contingency*. Two types of reserve are commonly used – cost reserve and 'time' reserve. A reserve is an extra amount (in money or time period) added to an initial cost or duration estimate respectively to take care of possible unknowns.

Resource

Resources refer to people and equipment/plant undertaking the production. They are distinguished from money which is a common unit of measurement of resource usage (Figures 3.42 and 3.43). They are also distinguished from materials which refer to that which is incorporated into whatever is being produced; materials include parts and components.

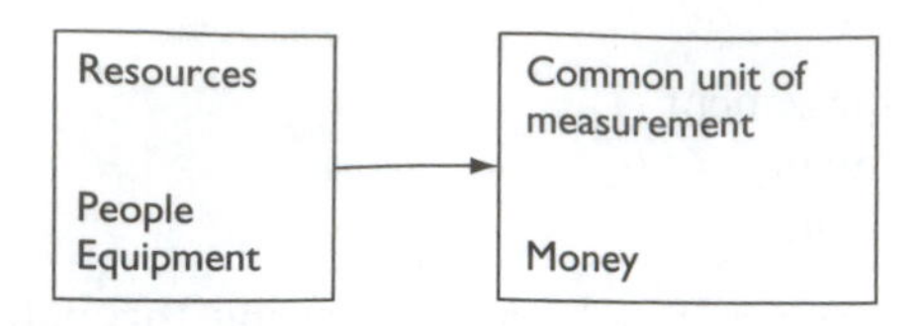

Figure 3.42 Preferred reference to resources.

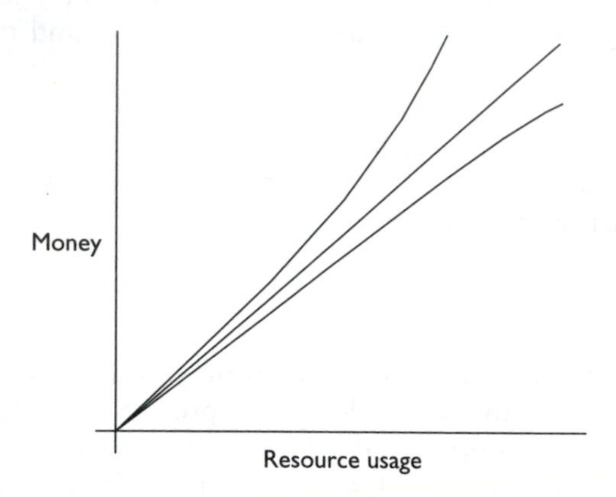

Figure 3.43 Some possible relationships between resource usage and money.

In this book, the terminology 'resources' may be used to refer to resources generally, to resource numbers (quantity) or resource types. The term 'resource types' is used where a distinction between the various resources is considered necessary. The term 'resource number (quantity)' is used where a distinction within a resource type is considered necessary.

Some people refer (inappropriately in planning) to time, overheads, skills etc. as resources.

Resource balancing

'Resource balancing' is a term applied to resourcing work, and typically resourcing sequential/repetitive work. Two extremes in handling resources are possible:

1. Resource numbers (quantity) are chosen to give a desired activity production rate. Project duration is considered a higher priority than project cost. Where multiple activities occur (sequentially or otherwise), it may be desirable to have the same activity production rates for each activity; under such circumstances, plots for each activity on a cumulative production plot are parallel, with or without a buffer between them (Figure 3.44).

2. The production rate for each activity is determined by the available or usual resourcing numbers (quantity). Project cost is considered a higher priority than project duration. Where multiple activities occur (sequentially or otherwise), plots for each activity on a cumulative production plot are not parallel; the start or finish times of activities are adjusted in order to avoid interference between activities, and the project completion date will relate to the lowest activity production rate (Figure 3.45).

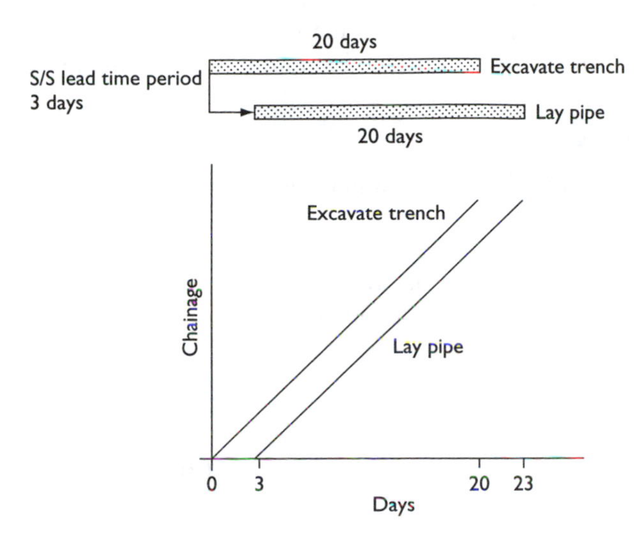

Figure 3.44 Example activities advancing at the same rate, and S/S relationship.

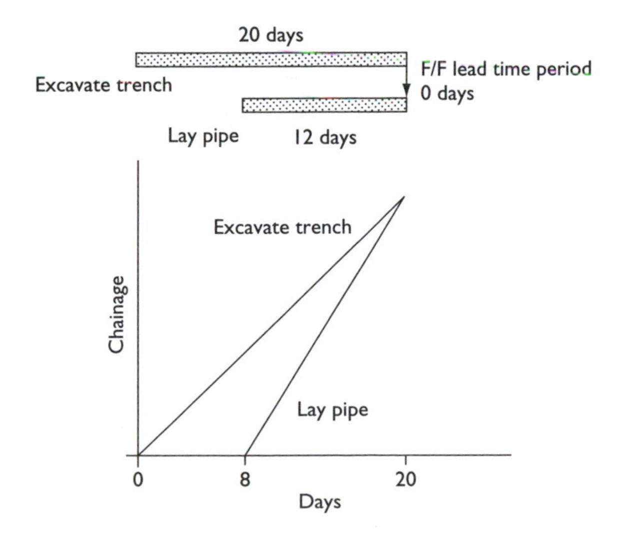

Figure 3.45 Example activities advancing at different rates, and F/F relationship.

Constituent level and element level controls are adjusted to give desired activity and project level performance.

Collectively, this thinking may be called the line of balance technique. The first method of handling resources mentioned above might be termed 'parallel scheduling'. The second method might be termed 'resource scheduling'.

Many projects have the same sequential work repeating throughout the project lifetime. An example is the construction of multistorey buildings where the same trades (for example involving the sequence of plumber, electrician, ceiling fixer, and painter) are repeated floor by floor. Other examples include pipeline and railway construction and estate housing projects.

Resource planning

Resource planning refers to that involved in deciding the usage of resources on a project.

The resource planning problem is a subproblem of the total planning problem, where emphasis is placed on the different level controls that can influence resource usage, and all other controls are held fixed.

Resource plot

Also called a resource profile or resource histogram. A resource plot gives resource requirements versus time, over the duration of the project.

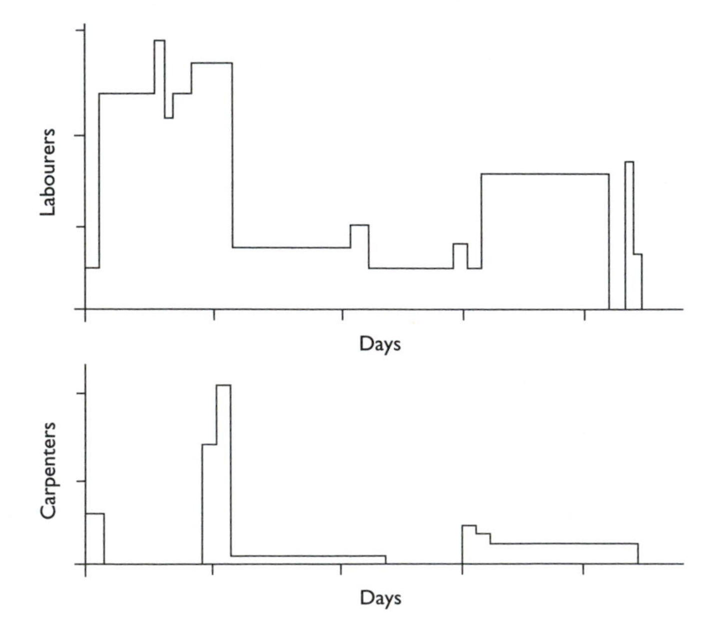

Figure 3.46 Example resource plots.

The resource plot is developed for each resource type by summing up the daily (or whatever time unit is applicable) resource requirements for each activity, as scheduled in a bar chart. This gives the total resource requirements on any day. Plotting these values gives diagrams like Figure 3.46.

The diagrams in Figure 3.46 are project level representations. Both resource smoothing and resource constrained scheduling practices use the resource plot as their starting point.

Resource smoothing; resource constrained scheduling

Problems associated with resource usage may be categorised as:

- Resource smoothing (also called levelling)
- Resource constrained scheduling.

The resource smoothing problem and the resource constrained scheduling problem are subproblems of the total planning problem, where emphasis is placed on the different level controls that can influence resource usage, and all other controls are held fixed.

Generally, when activities are conducted simultaneously, then this leads to simultaneous demand for resources, producing peak resource demands at certain stages of the project. Peak demands of resources, particularly over short time periods, may be undesirable.

Materials problems may be similar to those for resources, and may be handled similarly.

Resource smoothing involves evening out the resource requirements over the project or over stages of the project, by reducing peak demands for resource numbers (quantity) and creating a requirement for resources at other non-peak time periods. Smoothed resource numbers are desirable, for example, if people are the resource, because it can provide continuity in the workforce, eliminate undesirable hiring and firing and attendant industrial and people problems. Similarly it is not desirable to have plant and equipment being used intermittently (Figure 3.47a).

Smoothing the requirements for plant leads to higher utilisation and hence more efficient usage of these investments. Peaks in material requirements are to be avoided if only to avoid the possibility of shortfalls in supply during these peak time periods.

Resource constrained scheduling involves scheduling activities such that their resource requirements never exceed available numbers (quantity). For example, if plant or equipment is the resource, then large numbers of equipment may not be available. Some resource types can be expensive and limited in number (quantity). Skilled labour is often difficult to obtain and expensive to fire. That is, typical constraints will relate to upper limits on

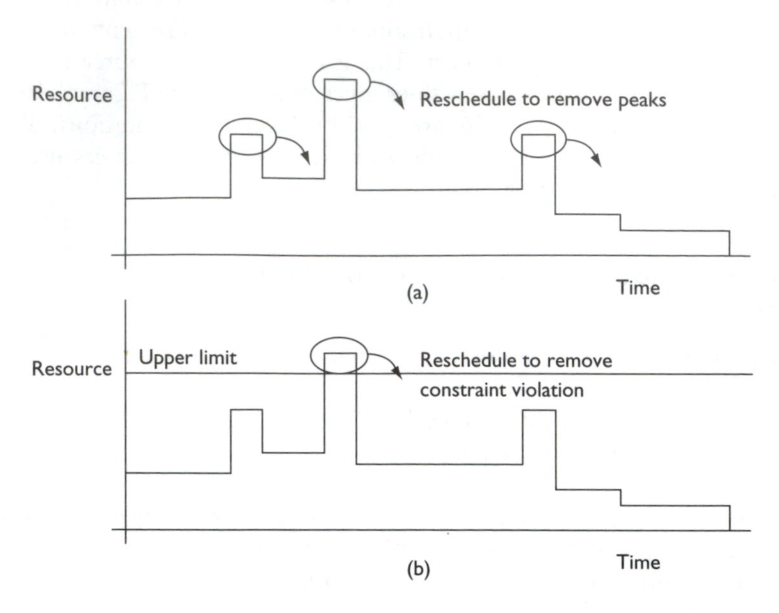

Figure 3.47 Schematic of what happens in (a) resource smoothing and (b) resource constrained scheduling.

Resources are taken into consideration when establishing the nature and order of work. No further treatment of resources is necessary.	Resources are ignored (or resources are assumed to be available) when establishing the nature and order of work. It is then necessary to carry out resource manipulation (resource smoothing or resource constrained scheduling).

Figure 3.48 Two broad approaches to dealing with resources.

people with certain skills or upper limits on numbers of pieces of equipment. A schedule is clearly not feasible if it requires resource numbers in excess of those available (Figure 3.47b).

Resource smoothing and resource constrained scheduling may or may not be significant problems depending on the degree to which resources are considered in the initial thinking. When planning is carried out in an iterative analysis fashion, two broad approaches may be observed when dealing with resources (Figure 3.48).

Most people seem to do something in between these two extremes. That is, they consider resources to a certain extent (maybe in their heads) up front in planning, but also possibly fine tune later on. The chosen work method itself can act like a constraint on the way resources are handled.

The solutions to both the resource smoothing and resource constrained scheduling problems are similar, if not always clear cut. Effectively, if the resource plot is thought of as hills and valleys, an earthmoving exercise removes the hills and puts the earth into the valleys. In more fundamental terms, the project controls (method, resources and resource production rates) are selected to give the desired resource usage over the project duration. That is, the solution steps to resource smoothing and resource constrained scheduling problems are no different to any project planning problem.

The solution could be expected to be more complicated when dealing with *multiple resource types* simultaneously.

Controls

A common first approach attempting to solve resource-related problems (in an iterative analysis sense) is to use activity level controls. If activity level controls don't work or don't work completely, then constituent level, element level and project level approaches may be tried. Of course, controls can be tried in any order the planner wishes, rather than following any suggested order. Some example controls follow:

Solution (controls) at the constituent level. Change resource productions through, for example, incentive schemes, bonuses, carrots and sticks. These feed into higher level solutions.

Solution (controls) at the element level. Change resource numbers (quantity). Shift work or overtime might be tried. These feed into higher level solutions.

Solution (controls) at the activity level. A control which is often tried first is:

- Make use of any activity float. The start dates of non-critical activities are moved. Note that shifting the start dates of critical activities will extend the project completion date.

If this doesn't work or doesn't work completely, then other activity level approaches may be tried:

- Try compressing the durations of certain activities. Compressing the durations of activities involves shortening the activity durations, but this usually comes with a penalty of increased cost and perhaps an additional and more concentrated resource requirement.

- Try lengthening activities, though this may come with a cost penalty, sometimes called uneconomical drag out.
- Try splitting activities that can be split. Splitting implies starting an activity, stopping the activity, starting the activity again after a break and so on.

Solution (controls) at the project level. Example project level controls are:

- Extend the project duration. All activities then become non-critical.
- Examine alternative methods of work. This will lead to alternative network logic. Examples include outsourcing or contracting out some work instead of doing the work in-house, using machinery instead of labour, and prefabrication rather than *in situ* work.
- Try overlapping relationships between activities. In place of more usual finish-to-start (F/S) relationships, start-to-start (S/S), finish-to-finish (F/F), or start-to-finish (S/F) relationships might be possible.

Other activity level and project level approaches are only limited by the planner's imagination. Knowledge of the project and the industry will help develop alternative approaches.

Whether any of these approaches are possible will be determined by the project and the nature of the project work.

Algorithms

Many of the better computer packages that perform network analysis come with resource smoothing and resource constrained scheduling options. The packages generally have inbuilt *heuristic algorithms* for solving these resource problems. The heuristic algorithms try something (such as the solution approaches listed above) (usually in increments of a day, or other time unit), and if it produces an improvement then more of the same is tried; if no improvement is obtained, then something else is tried (Figure 3.49).

Within the solution approaches listed above, there is still a need for the algorithm to prioritise what is tried.

- Try something.
- If an improvement occurs, try more of the same.
- If no improvement occurs, try something else.

Figure 3.49 Heuristic logic of resource handling algorithms.

If resource usage and activities are to be manipulated, there is no definitive rule as to which activities should be manipulated and the order in which they should be manipulated. Activities may be given rankings or priorities at the discretion of the planner (or computer package) according to any or all of the following as well as others:

- Earliest start time EST (lowest or first occurring).
- Earliest finish time EFT (lowest or first occurring).
- Latest start time LST (lowest or first occurring).
- Latest finish time LFT (lowest or first occurring).
- Duration (shortest or longest).
- Free float and total float (least or most).
- Resource requirement (largest).
- Activity number (lowest).

The allocation of resource numbers is carried out according to the selected priorities. Where activities have equal priority then appeal is made to the next lower priority to distinguish between the activities.

The above suggestions are possibilities for establishing a heuristic algorithm for solving the resource smoothing and resource constrained scheduling problems. Other suggestions may be found in the technical literature.

Such approaches will generally mean that a satisfactory solution may be found but one which is not necessarily optimal. Computer solutions should also be checked for practicality, and so some form of solution that involves *interaction* between the planner and the computer package will perhaps lead to the better overall solutions. All heuristic algorithms require the planner to attach priorities and a ranking to activities with different attributes. Each computer package could be expected to give a different solution.

Small projects can often have their resources rescheduled by inspection.

Measures

How does a computer package, that performs resource smoothing and resource constrained scheduling, know that one resource plot is better than another? By human eye, one plot can be seen to be better than another, but a computer package cannot 'eye' the plots, rather some quantitative measure is needed for comparison purposes.

Rescheduling, whether through such algorithms mentioned above or by hand, leads to different resource plots. Different resource plots may be compared quantitatively through the following measures:

- Sum of squares (moment)
- Number of hirings and firings
- Peak requirement
- Resource utilisation.

A resource plot with a low sum of squares, a low number of hirings and firings, a low peak requirement and a high resource utilisation is preferred.

For a resource requirement on day k of $u(k)$ the sum of squares measure is given by

$$\text{SOS} = \sum_{\text{all } k} u(k)^2$$

(A sum of squares calculation is analogous to a statistical variance [standard deviation squared] [second moment] style calculation. Compare the shape of a histogram or probability mass/density function with a high variance to one with a low variance. Compare the moment of inertia or second moment of area of a structural member's cross section, for example an I-shaped cross section with that of a rectangular cross section.)

The number of hirings and firings is the sum of the vertical portions of the resource plot. The peak requirement is the maximum resource requirement. The resource utilisation is the ratio of the area under the resource plot to the area under a rectangle drawn with one side as the peak resource requirement and the other side as the time axis.

Multiple resource types

When *multiple resource types* are present, weighted measures can be used. The weightings are chosen subjectively to reflect the relative importance of the resource types.

The weights take account of the relative magnitudes of the resource types. For example, the average requirements for two resource types might be 2 and 20 respectively; the weightings would be expected to reflect this difference in magnitudes.

A weighted sum of squares measure would look like,

$$\text{Weighted SOS} = w_1 \text{SOS}_1 + w_2 \text{SOS}_2 + \ldots$$

where

w_i weighting for resource type i, $1 = 1, 2, \ldots$
SOS_i sum of squares for resource type i, $1 = 1, 2, \ldots$

Similar weighted measures can be devised for the other measures of hirings/firings, resource utilisation and peak requirement.

The final resource plots depend on the values chosen for the weighting coefficients. It is only the relative values of the weighting coefficients which are important, not their absolute values. The final solution is a compromise solution between what is best for each resource considered singly.

Some authors suggest that the weighting coefficients be chosen so as to give a minimum cost solution, cost here being defined for each resource type as a unit resource cost (resource cost/day) plus a hiring/firing cost and the total cost being a sum over all resource types. However, this cost is problem dependent and generalisations from one problem to another are difficult to make. Hence to adopt such an approach for obtaining the weighting coefficients, a whole range of combinations of weighting coefficients has to be tried and the resource types smoothed and the cost evaluated for each case. The combination or combinations of weighting coefficients yielding the least cost solution is/are then used. Restricting the coefficients to integer values decreases the number of combinations that have to be searched. However, the number of combinations is still very large.

General

In general, the measures (for example, sum of squares) resulting from multiple resource types may be treated in several ways:

- The measures for all the resource types are combined into a single measure. The above use of weightings is an example of this.
- One resource measure is regarded as more important than the others, and the less important measures are converted to constraints.
- A solution is obtained for each resource measure separately, and these are then traded off against each other.

Such approaches are no more than is done in multicriteria (multiobjective) decision making or optimisation (Carmichael, 2004).

Responsibility matrix

Alternatively called responsibility chart. For each activity, there are attendant resource requirements. People resource and organisational (including coordination) requirements, responsibilities and authorities can be represented in a responsibility matrix form which may have rows corresponding to the activities, columns corresponding to resource types and entries corresponding to responsibilities, or similar.

The matrices are a way of informing project participants of their roles and duties.

Risk

The exposure to the chance of occurrences of events adversely or favourably affecting the project/business/. . . as a consequence of uncertainty.

$$\text{Risk} = f(\text{Uncertainty of event, Potential loss/gain from event}).$$

Risk management

The risk management process generally follows Figure 3.50.

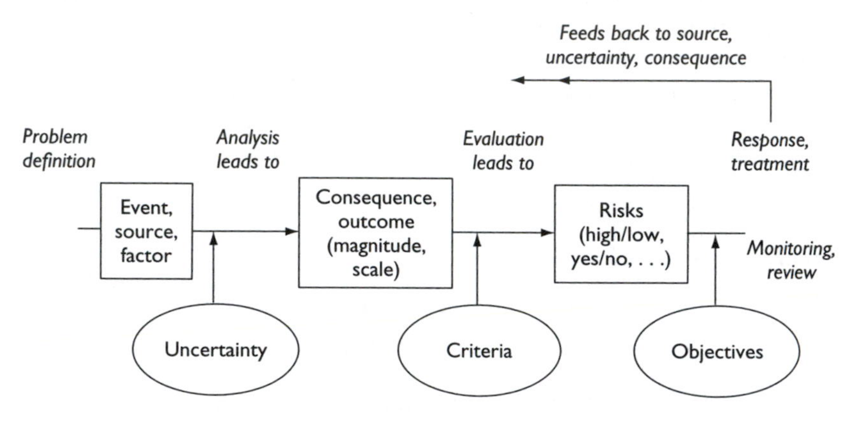

Figure 3.50 The risk management process.

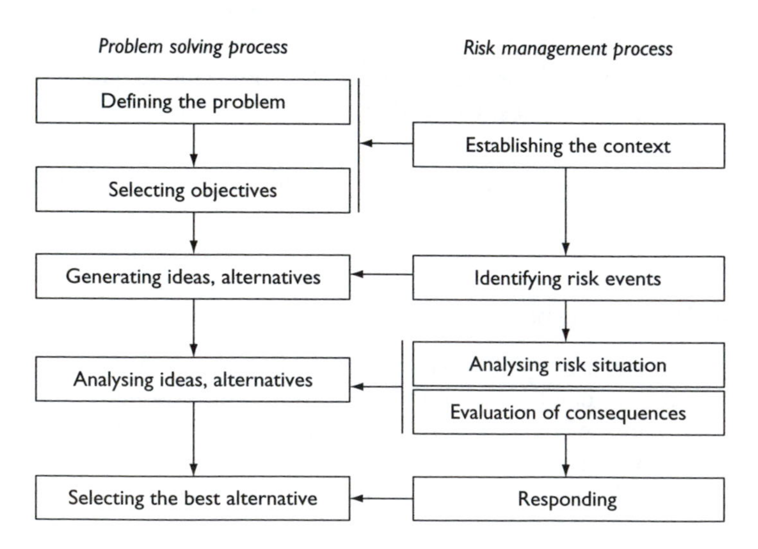

Figure 3.51 Problem-solving process and risk management process compared (Carmichael, 2004).

The risk management process is essentially the same as that used in systematic problem solving, though different terminology might be used and the number of step classifications might be different. Refer Carmichael (2004) and Figure 3.51.

S curve

An S curve is a cumulative representation of either a resource (people or equipment) or money over the project duration. The curve shape is like a stylised S for large projects (Figure 3.52). For small projects, the S shape may not be readily apparent. The S curve is an industry-accepted representation at the project level. The flatter portions correspond to the project starting up and the project winding down, with the section in between corresponding to 'steady state' progress of the project, where most resource and money usage occurs.

The basis for developing S curves is a bar chart. From the bar chart the resource (and money) requirements can be summed (integrated) over all activities on a day-by-day basis (or other applicable time unit). These daily figures (for each resource type, called a resource plot, histogram or profile; for money, called an expenditure plot or diagram) are then accumulated, going from project start to project end, to give the S curve (Figure 3.53). (The cumulative plot is obtained from the resource plot or expenditure plot in the same way a cumulative distribution function is obtained from a probability mass function or probability density function in probability theory. Those people familiar with soil sieve analysis will also recognise the steps of getting to the cumulative plot.)

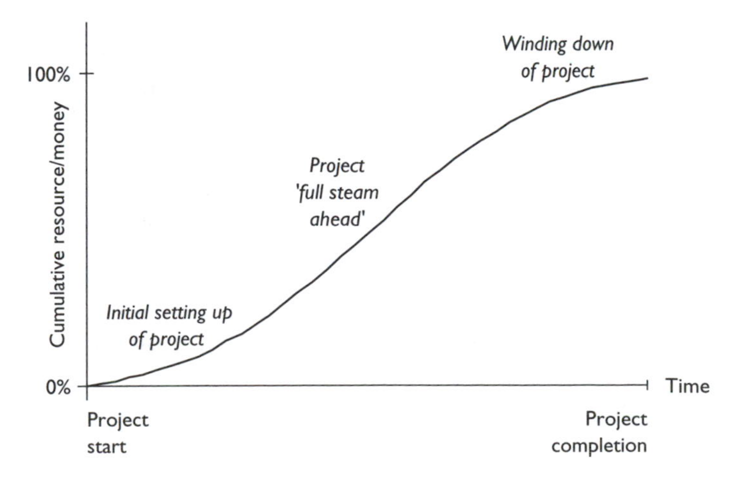

Figure 3.52 Example S curve.

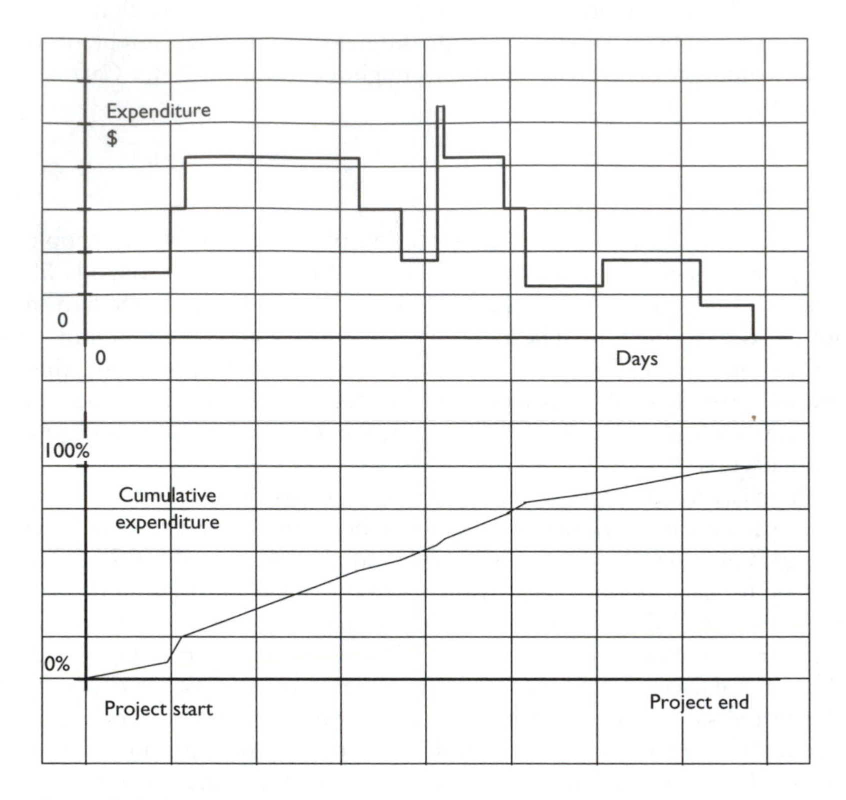

Figure 3.53 Example expenditure (upper diagram) and cumulative expenditure (lower diagram) plots.

Because the bar chart is not a unique entity for any project, for example, some activities have float and can be adjusted in their start times (between earliest start times EST and latest start times LST), so a number of S curves are possible for any project (Figure 3.54) for each resource type and money. For planning purposes, the particular S curve, which leads to the more agreeable resource and financial commitments, would generally be preferred and could be expected to lie somewhere between the two activity starting time extremes. The S curve for money may be referred to as a *cumulative expenditure plot* or *cumulative cost plot*. It is an industry-accepted way of representing planned expenditure at the project level, as opposed to an expenditure plot or diagram that is equivalent to a resource plot in presentation.

Cumulative expenditure plots are repeatable to a reasonable accuracy between similar projects, and may be used for *budgeting* purposes on second and subsequent similar projects.

The S curve also finds use in reporting for replanning purposes.

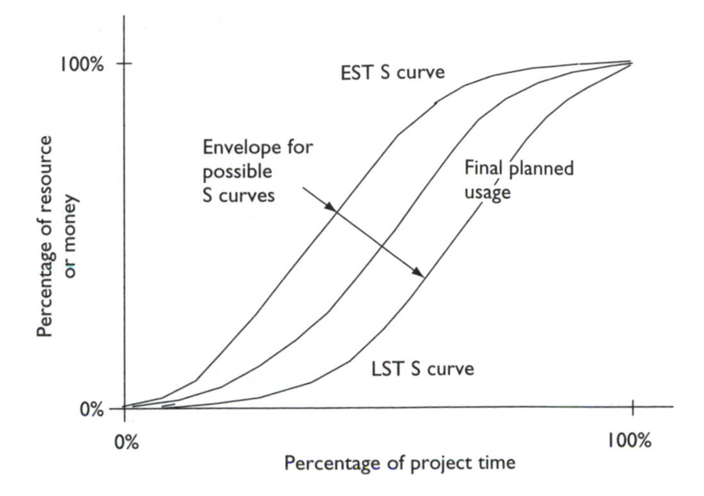

Figure 3.54 Envelope of S curves based on activity start times.

Schedule

Refer 'program'.

Scheduling

Establishing the timing and order of work, that is when work is to be done (and hence it follows, when resources and money are needed), is usually referred to as scheduling or programming. However, some people enlarge upon this meaning and loosely use the term interchangeably with planning.

Scope

Scope is what is involved in undertaking the project, the extent of the work to be undertaken, what work is contained in the project (perhaps by defining what work is not included in the project, in effect the boundaries to the project) that leads to the end-product. Scope is fully described by listing all the activities.

There is much confusion over the usage of the term 'scope'. Some writers wrongly include extra things besides project work. While useful information to have, it is not scope, and should be clearly distinguished from scope. The biggest transgression though is the sloppy, interchangeable use of the terms 'objective', 'constraint' and 'scope'; few people appreciate the distinction and why there is a distinction. Commonly the carriage is put before the horse, such that the scope is said to determine the objectives and the constraints. This confusion and imprecision stifles the understanding of project management.

Refer Carmichael (2004).

Stage

A stage corresponds to a subinterval of the project duration. It may have physical interpretation or not.

State

The state describes internal system behaviour. For the stripped-back projects considered in this book, output and state are the same, that is there are no observability (in the sense of Kalman) issues. State contains multiple parts (that is, it is a vector quantity of state variables). State is a controlled variable.

Task

Refer 'activity'.

Time management

The term 'project time management' (that is, the management of anything involving time on a project, and including program management) is not favoured in this book. There also exists the term 'personal time management' relating to how people 'use their time'. Neither form actually manages time, but the terminology is widespread. Also, some people loosely refer to planning and time management synonymously.

In this book, time is viewed as the independent variable, and projects are regarded as dynamic systems because of this. The term 'time' is used correctly when used in a technical sense, but is used in a lay person's sense elsewhere.

Time-chainage chart

Refer 'cumulative production plot'.

Time-scaled network

A time-scaled network is an activity-on-link network drawn to a horizontal timescale. There is no vertical scale. The lengths of the activities (links) are made proportional to the activities' durations. A time-scaled network gives the same information as a connected bar chart, namely when activities begin, when activities end, how long activities take, the amount of float in activities, and the dependence between activities. The horizontal scale is ideally matched to a calendar allowing for stipulated work hours per day, weekends, public holidays, rostered days off (RDO), number of shifts per day and so on.

A time-scaled network may be referred to as a program or a schedule.

A time-scaled network follows directly from a network analysis. The information from a network analysis is readily transferable into a time-scaled network form.

Figure 3.55 gives an example time-scaled network.

As with networks and bar charts, different drafting styles are encountered. For example, the vertical lines in Figure 3.55 might be drawn inclined. Planners use whichever style appeals to them.

In time-scaled networks, activity durations are given by the horizontal components of lines; the vertical components have no consequence. Float is usually indicated by lines with texture different to that used for the scheduled duration of activities.

Only activity-on-link networks can be drawn to a timescale. Activity-on-node networks do not lend themselves directly to the time-scaled version. However this does not stop some people from pretending to draw activity-on-node diagrams to a timescale. Typically such diagrams end up looking like Figure 3.56.

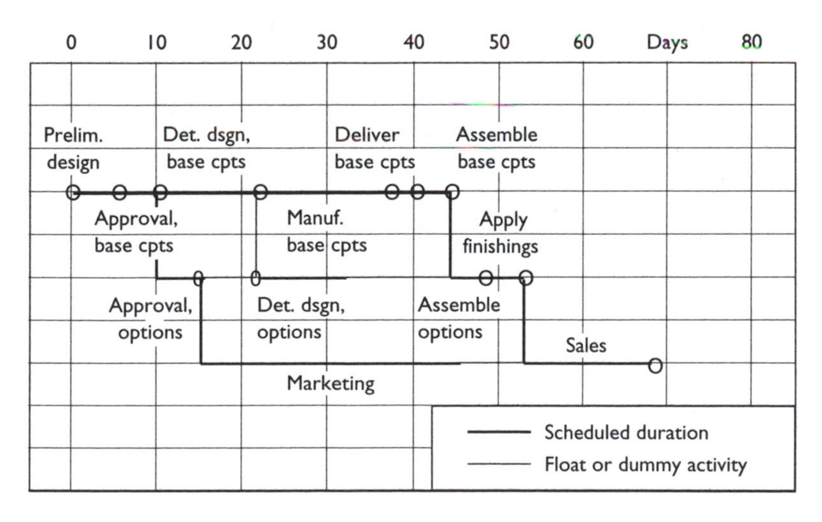

Figure 3.55 Time-scaled network for an example project. See the associated bar chart, Figure 3.1 and networks, Figure 3.25(a,b). (Activity names have been abbreviated because of space limitations – work items are implied.)

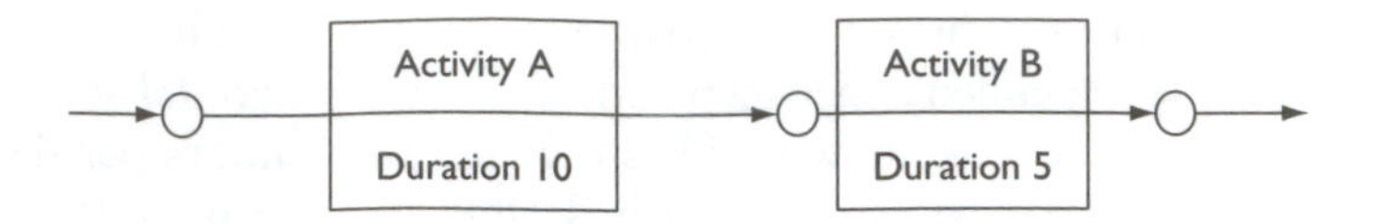

Figure 3.56 Pretend time-scaled activity-on-node diagram.

A true time-scaled activity-on-node diagram would have elongated and contracted nodes (circles, rectangles, . . .), and would look silly. Instead, diagrams like Figure 3.56 are drawn, but these are activity-on-link diagrams disguised to look like activity-on-node diagrams.

So, when a planner uses a connected bar chart or a time-scaled network, the planner is unwittingly using an activity-on-link diagram representation, even though that planner might confess to being a staunch advocate of, or show a biased preference for, activity-on-node diagrams over activity-on-link diagrams.

Time-scaled networks also find use in reporting for replanning purposes.

Timetable

A timetable is a document showing when events occur. It may be referred to as a program or a schedule.

Value management/analysis/engineering

Value management (equivalently value analysis, value engineering) is popular in project management for a diverse range of situations. Originally conceived as a way of economising on resources or finding alternatives, it finds application at all phases of a project from concept through to termination. Central to the approach is a systematic analysis of function and an examination of alternatives. As such it can be shown to be but a special case of systems engineering/problem-solving methodology.

Its origins trace back to about the 1950s when it was known as *value analysis* or *value engineering*. Then it was a design review or 'second look' approach to proposed or existing designs. Its area of application has enlarged over the intervening years such that it now encompasses not only design reviews but also, for example, feasibility studies, an examination of project goals, and conflict situations.

Constructability or *buildability* studies are but special cases of value management applied to construction or building projects.

Central to value management is the analysis of function from a whole system viewpoint and the proposing or generating of alternatives. It identifies wastage, duplication and unnecessary expenditure, and can assist in testing assumptions and needs. The whole system or holistic approach avoids conventional compartmentalised thinking and obtaining locally optimal or suboptimal solutions at the expense of the desirable globally optimum solution.

Refer Carmichael (2004).

Variance

A variance is the difference between planned (or as-planned) and actual (or as-executed, as-built, as-constructed, as-made, . . .). It is not referring to any probabilistic measure of variability.

$$\text{Variance} = \text{Planned} - \text{Actual}$$

The usual convention is for a variance to be negative if a project is behind production/schedule or, for example, over budget. A variance is positive if the project is ahead of production/schedule or, for example, under budget.

Using earned value, several types of variances can be identified, including cost variance and production/schedule variance. There, special definitions of variance are used.

Work breakdown

A project (scope) can be decomposed into subprojects and then into sub-subprojects and so on down to the activity level. The lower levels represent finer detail. The process of decomposing a project (scope) might be called scope delineation (or scope definition) by some writers. The term *work breakdown structure (WBS)* is used to refer to the resulting systematic decomposition (as opposed to an unstructured activity list where there is a possibility that some activities have not been thought of).

Work breakdown is not unique; various work breakdowns are possible for any project. A breakdown is chosen that makes life as easy as possible for the planner. The degree of coarseness or fineness with which a project is subdivided will depend on the end use of the plan. There is no right or wrong answer to this, only better or worse subdivisions.

Some people (wrongly) see this breaking down as going from the 'product to the processes', or as a product-oriented tree. However, since a project is not a 'product', what is happening is more akin to generally defined processes being broken down into more refined processes.

Some writers (wrongly) say that scope delineation involves subdividing the project deliverables. (Deliverables are the end-product plus project state final conditions.) But only project work should be involved in the breakdown, and not the end-product.

Planners distinguish between events and activities. Each activity has events associated with it. Planning with either is okay, but care has to be exercised when the two are mixed. Some writers (wrongly) include events in work breakdown structures.

Terminology for each of the levels, used by different people, might include project, subproject, work package, job, subjob, process or activity. The terminology of task and subtask might also be used at the middle to lower levels. There is no consensus in the use of terminology to describe the various levels, though this book's preference is for the terminology *project – subproject – ... – activity*. 'Program' is a term used to refer to the level above projects and represents a collection of projects; that is a project is a subsystem of a 'program'. An alternative usage of the term 'program' is given above.

Work method

Work method is the sequence or logic of operations, with activity and event interrelationships. It is commonly represented by a network.

Work study

Work study attempts to create in people a questioning state of mind in the way they view their current and future work practices. This aware state of mind applies to all phases of a project and on a continuing basis in an organisation. Work study involves the critical and systematic analysis of work with an ultimate view to improvement and the elimination of any nonproductive components. Work study is made up of *method study* and *work measurement*. Method study breaks down work into its components and questions the purpose and need of each component; it involves, amongst other things, recording information on the work, critically examining the facts and the sequence and developing alternatives. Work measurement is concerned with the time periods taken to perform work. Work study is an established technique that has wide applicability.

Work study may be broadly defined as an examination of the use of people, equipment and materials in work tasks, and an associated attempt at improvement and elimination of waste. Besides the financial incentives, it attempts to create an attitude of mind about the effective use of people, equipment and materials.

Work study can be shown to be a special case of systems engineering/problem-solving methodology. Re-engineering is work study reinvented (Carmichael, 2004).

Exercises

1. Under what circumstances should incomes and expenditures, as shown in Figure 3.17, be plotted based on dates of anticipated transfer of funds and billings, or based on invoiced, committed or paid amounts, or other?

2. (a) Isolate the parts of the 'classical' management function approach of what are categorised as 'planning', organising, staffing, directing, 'controlling' and coordinating that imply involvement with (i) schedule/production, (ii) money, and (iii) resources. How might this 'classical' management function division help or hinder the understanding of planning as outlined in this book?
   (b) Isolate the parts of the management function approach of scope management, quality management, 'time' management, cost management, risk management, contract/procurement management, human resources management and communication management that imply involvement with (i) schedule/production, (ii) money and (iii) resources. How might this management function division help or hinder the understanding of planning?

3. Some planners include in their program an allowance for delays – due to weather, industrial action or other unforeseen circumstances. These delays are incorporated into programs as separate activities. The argument behind this is that the overall program should present as realistic as possible picture of the duration of the project.

What do you think of such a practice?

Some planners will include delays in all programs except at the very detailed level; at the lowest level it is believed that by incorporating delays, focus on target dates disappears. By including delays does it take away the emphasis on target dates?

Will knowledge that an allowance for delays has been included in the program make people think that delays are inevitable or expected? Is this a case of Parkinson's Law? – *Work expands so as to fill the time available for its completion.* Is this a case which supports the notion of self-fulfilling prophecies?

Will some of the project team assume they have longer to do their work because a delay has been included in the program for other members of the project team – for example, will designers take longer knowing that a delay has been incorporated into the site work program?

Should delays only be incorporated in the broad program, with detailed programs omitting delays in order to focus attention on target dates?

Should there be two programs – one without anticipated delays (a target plan, made public) and the other with delays (kept private)?

4. Obtain a user's manual for a computer program that does network analysis including resource handling and overlapping relationships. Research the program input and program reporting facilities.

5. For the following project network in activity-on-link diagram form, convert it to an activity-on-node diagram form. For convenience, activities have been listed as letters of the alphabet.

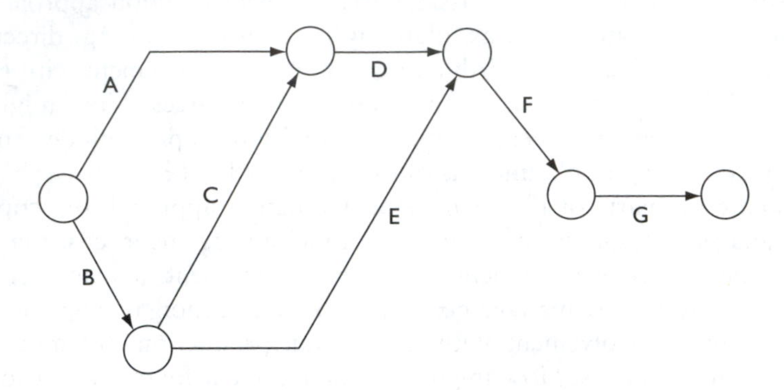

Convert the following project network in activity-on-node diagram form to an activity-on-link diagram form.

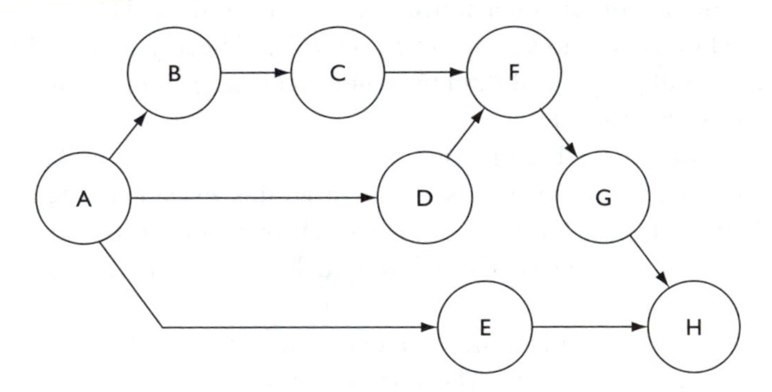

6. Why might some planners dogmatically criticise activity-on-link diagrams (while stating a preference for activity-on-node diagrams), yet quite happily use bar charts?

7. An S curve (money) represents cumulative expenditure. Why is this form of diagram preferred by industry over reporting on individual activity expenditure, much like a bar chart reports on individual activity durations?

8. In examining some proposed forestry (timber harvesting) operations, an environmental impact statement is prepared. The main activities in this are listed in Table E3.1. The order in which the work is planned to be done is according to the network logic of Figure E3.1.

Table E3.1 Forestry study activities

Activity code	*Activity*	*Normal duration (days)*
Object	Establish proposal objectives and need	10
Descrip	Develop description of proposal	3
Altern	Examine alternatives	2
Physic	Study physical environment and the effects of forestry activity	20
Atmos	Study atmospheric environment and potential impacts	6
Hydrol	Study hydrological environment and impacts of the proposal	15
Veget	Study vegetation and effects of the forestry activity	20
Fauna	Study the fauna and effects of the forestry activity	20
Econo	Examine economics and land-use effects	15
Cultur	Examine cultural environment and effects of the forestry activity	10
Coord	Coordinate project	25
Report	Produce report	55

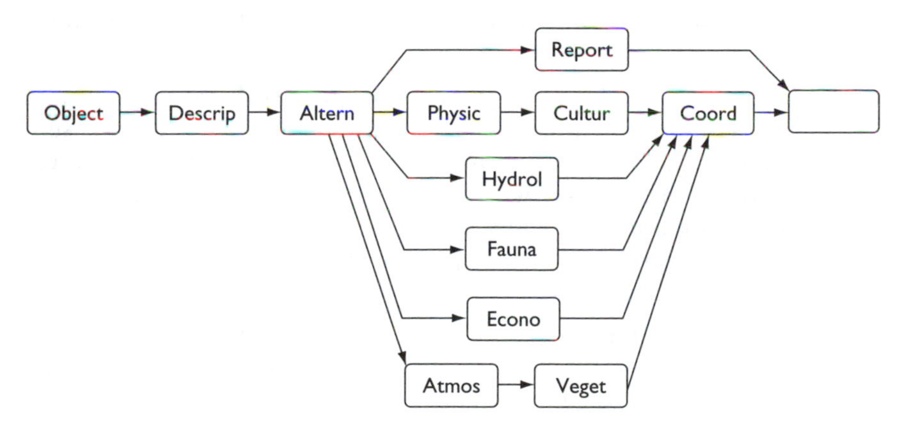

Figure E3.1 Forestry study logic.

Assume each of the 'Study' activities may be reduced in duration by 20% and the remaining activities by 10% (each rounded up to the nearest day). What is the possible shortest duration?

Chapter 4

The planning process

Introduction

Planning dimensions

Planning starts with broad assumptions which are steadily refined, as more information comes to hand. Planning typically proceeds from the broad to the detailed.

Planning is typically done from the general to the particular, from coarse to fine, in two directions (or two dimensions) (Figure 4.1) at the same time:

- In hierarchical *levels* involving breaking (the work in) projects into subprojects (for subprojects other than that based on stages), sub-subprojects, . . . , activities, elements and constituents. Detail emerges with lower levels. Project work is dissected into manageable portions through work breakdown (discussed later).
- Over *time*. Detail evolves as the project progresses.

The launching point for the main project planning is at the time of establishing the initial scope (Figure 4.1), which derives from the objectives and constraints. For each and all levels and at any time in Figure 4.1, a planning problem exists.

A plan near the top left of Figure 4.1 might be called a 'project preliminary plan' or similar name. It contains broad information on work method, resources, dates and so on. And it may be used as a basis for approval to do further project work, as might other plans at later stages.

As a synthesis problem

Planning is a synthesis problem. As such, there are multiple solutions (choices of control) possible. In most cases, planners are only after a satisfactory solution, or a solution that they can live with, and do not put

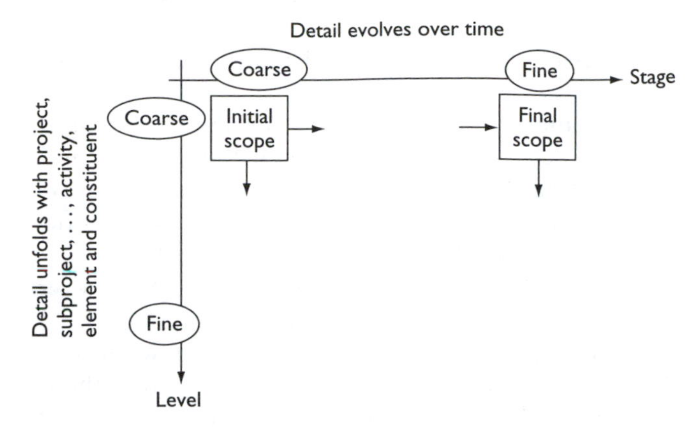

Figure 4.1 Evolution of planning problems.

additional effort into searching for the optimal solution. A planner may also be under pressure to come up with quick solutions.

Planners are unaware of and do not understand the components of the synthesis problem, and so they never know where they are relative to the optimum. They are unable to vocalise or formulate the synthesis problem components. Such discussion goes to the very heart of current planning knowledge being in its infancy, and current planning education (read training) being superficial and low standard.

As an iterative analysis problem

The solution of any planning problem via an iterative analysis mode involves the conventional steps in systematic problem solving or systems engineering.

Feedback occurs between steps in trying to refine the problem. (This is additional to the feedback that can occur between levels.)

Alternatives might be arrived at through the generation techniques used in a *value management/analysis/engineering study*, which aims at developing alternatives that perform a desired function but at a lower cost, or in a *work study*, which is concerned with the method and timing of work activities and their optimisation. Both value management and work study are no more than versions of systematic problem solving (Carmichael, 2004).

Case example – Small commercial buildings

This case example looks at one contractor's procedures for planning the construction of small commercial buildings involving minor earthworks, concrete work, stormwater drainage, steel fabrication/erection and electrical fitout.

Three phases/types of planning – initial, detailed and ongoing – are used.

The initial or preliminary planning occurs during preparation of a tender. A program is developed for submission with the tender and to assist the estimating department in pricing the project; the program may be required to assist the owner to further assess the contractor's offer.

The project duration may be set by the owner and it is a case of allocating sufficient duration to each activity so that the overall project is completed within the nominated duration.

The detailed planning does not occur until the tender has been successful. When a contract has been awarded, and within the first couple of weeks, a detailed construction program is developed. It is commonly a contract requirement that a construction program be submitted within 14 or 21 or 28 days of contract commencement.

Work on the construction program incorporates a review of the initial program, specifications and drawings. After possible discussions with the site foreman, it is determined which activities of the initial program can be retained and which activities require further breaking down. The aim is to develop a program with sufficient detail that can be easily followed and updated. Experience helps in determining activity durations and overlapping relationships to complete the project in the nominated time frame. This information is then input to a commercial project management computer package so that the critical paths, floats etc. can be determined and so as to have the program in a professional form for submission. The program is presented as a connected bar chart.

Ongoing planning of the project leads to updating the construction program on a monthly basis and the preparation of fortnightly programs (programs for the following two weeks). The fortnightly programs contain considerably more detail than the construction program and are usually updated on a weekly basis.

Planning problem components

The main components of any (optimal) synthesis problem (Figure 4.2), to which planning belongs, are (Carmichael, 1981):

- Model
- Objective(s)
- Constraints.

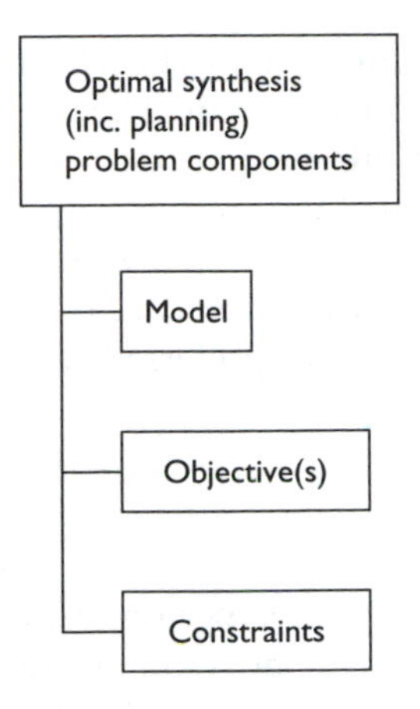

Figure 4.2 Optimal synthesis and planning problem components.

Objectives and constraints are developed below. Models are developed separately (Figure 4.3):

- Hierarchical models – Chapter 5, Chapter 9
- Multistage models – Chapter 8. The project is broken down into stages, or equivalently the project duration is discretised
- Probabilistic models – Chapter 7.

These problem components are expressed in terms of control, state and output variables. The system notions of control, state and output variables are central to the development of models and thinking for planning purposes.

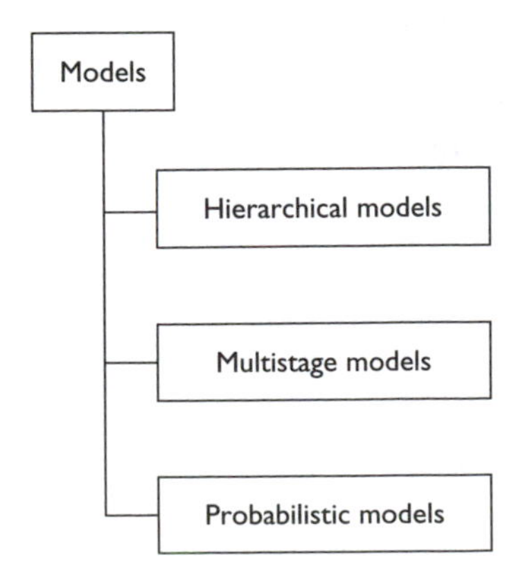

Figure 4.3 Model development in this book.

Control

Control is equivalent to a planning decision, action or input. The term 'control' is favoured here in place of decision, action and input. The planner exerts control on the project in order to bring about a desired project behaviour. The most obvious controls are work method, resource (people and equipment) selections and resource production rates. Work method, resource selections and resource production rates may be freely selected by the planner, subject to any constraints being present. The preferred controls are those that extremise the objectives.

State, output

The output describes external or observable response, performance or behaviour. Internal behaviour is usually given in terms of system state. By the nature of projects (in the stripped-back form treated in this book), the internal and external behaviour are the same. (The observation equation of control systems theory [Carmichael, 1981] becomes 'output = state'.) There is a one-to-one transformation between output and state. The terms 'output' and 'state' become interchangeable when referring to projects, and there is no noise in the observation device. There are no observability (or controllability) (in the sense of Kalman) issues for projects thought of in a stripped-back form.

The most obvious states are the resources used so far (cumulative resources) or the cost so far (cumulative cost) and the production so far (cumulative production).

Disturbance

In most systems studies, there is something that prevents the hoped-for output being obtained exactly. Commonly this is because of a changing project environment brought about because of delays, disruptions, adverse climatic effects and so on. Collectively the corruption is referred to as disturbance below, though control texts may use the term 'noise' (Figure 4.4).

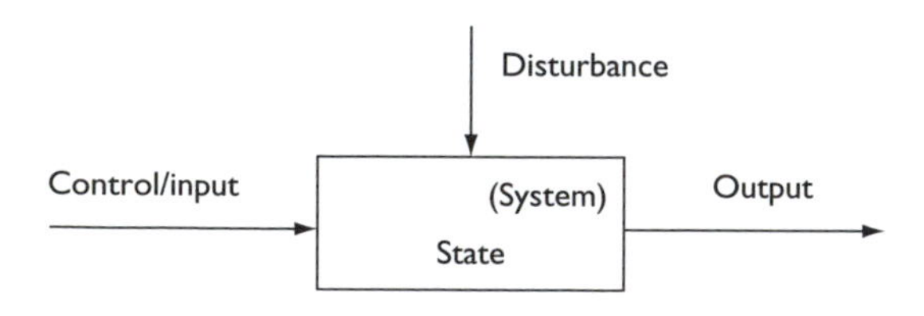

Figure 4.4 Schematic representation of control, state and disturbance.

Example

Planning establishes how and what work will be carried out, in what order and when and with what resources (additionally expressed in a money unit).

Consider the simpler (subsystem) problem where it has already been largely established how the work will be done and with what type and number of resources, and at what resource production rates. As such, the controls available to the planner are the order or sequence of the work including when particular work items will be done.

This establishes when resources are used, or, if expressed in a monetary unit, when money is used.

What the planner is trying to influence is the utilisation of the resources (state variables), and thereby the duration of the work.

There may be constraints on costs, duration of the work and resource availability.

Ultimately, what the planner might be trying to achieve is a work sequence of minimum cost or minimum duration or a compromise between these two objectives.

A suitable system model is the time-scaled critical path network or connected bar chart, with development to resource plots and cash-flow diagram.

As the planner adjusts the sequence of work, so resource utilisations, alternative cash flows and project durations evolve. By a series of iterations, the planner settles on one particular sequence of work.

Using an optimal control approach, the components to the planning problem are:

Model – Time-scaled network, connected bar chart, . . .
Constraints – Related to work duration, costs, resource availability and resource production rates.
Objective(s) – (Minimum) cost and/or (minimum) duration.

(After Carmichael, 2004)

Simplifications to planning

Ideally, project planning should be carried out as a probabilistic (inverse) synthesis problem. However, the tools are not currently available for this as a stand-alone mathematical problem. The full planning problem and its solution are very complicated, requiring ingredients of creativity and

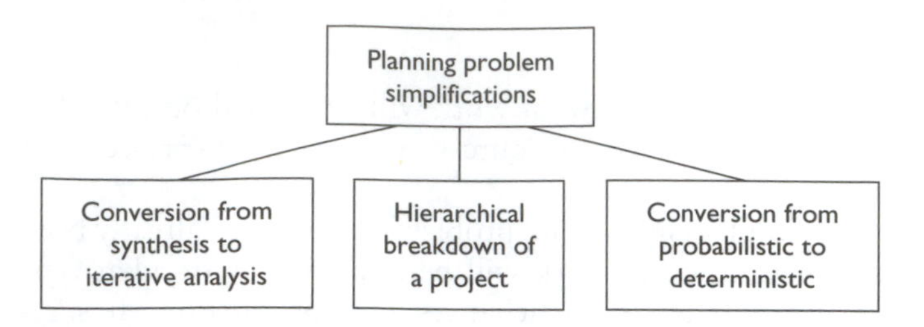

Figure 4.5 Simplifications to make the planning problem more tractable.

experience. Instead, planning is simplified and made more mechanical in three main pragmatic ways as shown in Figure 4.5.

- Instead of solving the planning problem in a synthesis format, it is converted into an iterative analysis format. Analysis is much easier than synthesis, and if a good initial guess is made, then there may be only one iteration (or even zero iterations) needed to give a sufficiently good answer. Iterations may modify the method, resource usage and resource production rates until satisfactory (without necessarily being optimum).
  For most planners, the step of generating alternatives is relatively straightforward without knowing how close to the optimum the guessed solution is.
  In some cases, where the planner is feeling his/her way, the analysis may become a 'what if' or sensitivity analysis examining the impact of alternative controls.
- The project (work) is broken down (hierarchically) to lower level subprojects, sub-subprojects, and so on to activities, elements and constituents. Subprojects may be chosen such that the interaction between subprojects is minimal and is commonly neglected, permitting subprojects to be planned independently of other subprojects. The same holds true where multiple projects are combined on an organisation level. However, interaction at levels lower than subproject is commonly strong.
  Interaction occurs because of the work method (implying an order/sequencing of work). Iterations (between levels) can largely be removed by the planner being aware of what is happening on other levels when solving an individual level problem. The levels are uncoupled as far as possible, for tractability reasons, to make the level problems less complicated.
- The planning problem is assumed deterministic. The analysis is made deterministic. Estimates for resources, productions, durations, costs, . . . are made deterministic in nature. Future events (leading to both upsides and downsides) are anticipated. Where future events have

a likelihood (uncertainty), this is incorporated into thinking akin to risk management; there some probabilistic analysis technique, such as Monte Carlo simulation, may be used, but more usually is handled deterministically through a 'what if' or sensitivity analysis.

Planning is always carried out over the full project time span. Planning stage by stage, without consideration of upcoming stages, could be expected to lead to a suboptimal solution.

Replanning

Replanning is carried out according to stages or at discrete points in time, but this follows from the nature of what replanning is. Replanning is no different in form to planning. The planning problem (for the remainder of the project) becomes better defined as the project progresses. Information assisting the planning problem could be expected to be better the less far into the future. Replanning is based on the remaining duration of the project using the current state of the project as initial conditions. A good plan allows for some flexibility and can be adapted to changing circumstances that may occur during a project. However, attempts are made to foresee potential influences in upcoming stages, and controls are chosen with these influences in mind. According to Machiavelli in The Prince: *When trouble is sensed well in advance it can easily be remedied; if you wait for it to show itself any medicine will be too late because the disease will have become incurable.*

Whether replanning is necessary or not is established through monitoring/sensing the project output ('as-executed') through time, and comparing this with the current baseline ('as-planned'). The difference between the 'as-planned' and 'as-executed' values (together with an updated view of future influences) may then be used to indicate what controls need to be modified, that is, to help in the generation of alternative controls for replanning.

Planning at some stages may take on increased importance for different stakeholders, for example for an owner at project outset, and for a contractor when tendering and while undertaking the work.

Objectives and constraints

General

Objectives and constraints derive from the value system of the project/end-product owner. Objectives and constraints lead to scope, and other planning matters. Scope is selected based on stated objectives and constraints. Without defined objectives and constraints, the ensuing scope and planning is ill-defined.

The term 'planning objectives' can occasionally be found in writings on planning, but its intent is not in a systematic problem-solving sense, and hence is not used here.

The objectives and constraints carry through to all planning problems at all levels (project, activity, element and constituent) and over time. They may be decomposed to finer detail and stated more specifically for use in lower levels, and later times.

Objectives

There will be an identified need or want for some product, facility, asset, service etc. This end-product is achieved through a project.

The materialisation of the end-product can be performed, possibly, in an infinite number of ways. In systems studies terms, the problem is an inverse problem, and hence there is no unique solution.

The criteria by which the preferred materialisation of the end-product is selected are the project objectives. Scope and other planning considerations on projects follow.

In a similar idea, there are possibly an infinite number of versions of end-products. The criteria by which the preferred end-product is selected are the end-product objectives. Form, function, finishes etc of the end-product follow. The selection of the preferred end-product form is a design problem.

That is, on any project there are two types of objectives

- End-product objectives
- Project objectives.

See Figure 4.6.

Commonly, project objectives say something about project cost, project [duration] and deviation from specification, but other objectives are possible. And these may apply throughout the project (for example minimum deviation from specification), or at the final (terminal) point of the project (for example minimum [duration which is related to the project] completion time). That is, general project objectives will contain a component over the time domain of the project and a component at the final (terminal) point.

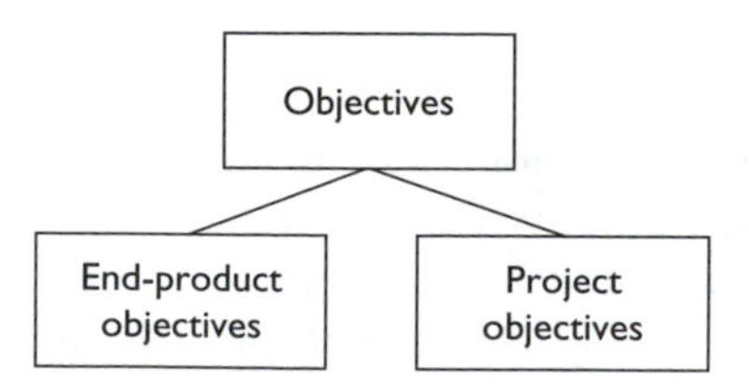

Figure 4.6 End-product and project objectives.

Example issues included in objectives

End-product objectives and project objectives (and constraints) may relate, for example, to:

- Money (end-product – sales, benefit:cost ratio (BCR), net present value (NPV), . . . ; project – cost, budget, . . .)
- Duration (end-product – lifetime, . . . ; project – duration, . . .)
- Resource usage
- Quality issues
- Community acceptance
- Environmental (natural) effects
- Safety
- Risks
- Public impact
- Extreme event impact (floods, cyclones, . . .)
- Social impacts
- Geotechnical considerations.

These are expressed in terms of end-product matters or project matters, as the case may be.

Confusion can arise when people start expressing project objectives in terms of the end-product requirements or end-product objectives.

For any situation there may only be a single objective or there may be many objectives (*multi-objective*). Where there are multiple objectives, they may be *non-commensurate* and *conflicting*. One or more of the objectives may be more dominant than the others.

The solution of such multi-objective problems generally involves some subjectivity while the solution of the single objective problem does not.

Constraints

All projects have genuine constraints such as funding, environmental (natural) and political. Constraints limit the range of controls possible; they restrict the options that are possible.

As with objectives, many people confuse a project constraint with an end-product constraint. End-product constraints may influence project constraints (Carmichael, 2004).

The higher levels and early project stages constrain the choices in the lower levels and later project stages respectively. Detailed planning is established within the parameters of previous broader planning.

Constraints can be converted to objectives (and vice versa).

Example objectives (project, activity, element and/or constituent levels)

- (Minimum) impact of activities on the natural environment
- (Minimum) project duration
- (Maximum) utilisation of resources
- (Maximum) return on money invested (in project and end-product) – related to usage of available money.

Example constraints (project, activity, element and/or constituent levels)

- Availability of resources, of the required type and in the required quantities when needed; a smoothed requirement for resources
- Availability of money, in the required quantities when needed
- A certain specified amount of interaction with the community
- Complete the work for less than (money). Cost. Budget. Spending limitations. Profitability. Cash flow
- Hazards and local factors; safety issues
- Statutory, legal requirements; environmental considerations
- Allowable sound volumes
- Site space restrictions; resource movement, storage, access, work flow
- A lower limit on the standard of work; quality assurance
- Designated milestones. Deadlines. Project completion time (may be imposed or as-planned)
- Production/schedule delays due to, for example, funding approvals, regulatory approvals, site access limitations, manpower availability problems and weather problems
- Events that are required to happen at the right time
- Technical knowledge, capabilities
- Transport, delivery issues.

It is frequently the case that a duration constraint or a cost constraint will be binding in the planning solution. That is, the preferred planning solution will lie on either the duration constraint or the cost constraint.

Care has to be exercised that any imposed constraint, especially an unrealistic cost constraint, does not prevent the free choice of sensible controls (method, resources and resource production rates). A certain desired standard/quality in performance may leave little flexibility in the choice of resources. In the extreme case, a constraint imposed without thought may be such that there are no feasible controls. Hence if the project proceeds in whatever form, the constraint will be violated; this may represent, for example, a financial loss or a completion time overrun, or the scope may have to change – 'you get what you pay for'.

Case example – Fruit processor upgrade

This case example discusses the practices used within a company to assign resources to a project involving an upgrade of an owner's existing fruit processing. The upgrade involved civil, structural, mechanical and electrical work, and the completion date was fixed by the need to have the work completed and the fruit processor running when the first fruit was ready for picking.

The company doing the project work was structured in three sections along discipline lines – civil/structural, process/mechanical and electrical/instrumentation.

Once the project manager had determined the resources required on the project, section heads assigned these resources to the project.

Following scope delineation, a program was developed to show how the work had to be done in order to meet the end date. The construction program was developed backwards based on the defined end date and the delivery dates for some long lead time period items that had been previously ordered.

The time period before the construction started then became the time period available for design and documentation, after allowing a sufficient time period for tendering. The other factor driving the design phase program was that the project manager was going to site for the construction phase and, in order to most efficiently run the design phase, all the documentation had to be completed before the project manager went to site. This resulted in a six-week time period in which to design and document the job.

At this stage, based on the scope (of work) for their sections, the section heads advised the number of hours they required to complete their work. They were given an indication of the program, and were asked to advise who they would assign to the job.

The hours given by the sections formed the basis of the fee proposal presented to the owner, and the owner instructed the company to proceed and commence design.

At this stage the project manager felt that all necessary resources were in place because all section heads had the hours they had asked for, and the project was proceeding on schedule.

As the design phase proceeded, it was found that the people who had been identified to work on the project were being diverted from this project onto other projects, and the design began to fall behind schedule.

Discussions with the section heads brought no solution and it was only after the project manager complained to one of the company directors that pressure was brought to bear on the section heads and the project was assigned the resources it needed.

At this point the design was about two weeks behind schedule. Fortunately, it was possible to get the critical documentation done during the period of reduced resource availability so that the construction schedule did not suffer greatly.

Off-setting this partly was a delay in receiving some information from the owner, which also delayed the documentation for a component of the work, and so the lack of resources available for this particular work package was not a problem.

The project encountered difficulties because it was only one of a large number of projects which were then happening, and which required resources from a finite pool.

Scheduling the use of resources could be difficult at the best of times, because owner approval for projects was often delayed, sometimes for weeks, and the company had to try and juggle the resources it had among whatever projects were going at any one time, and hope that the projects lying in the wings did not all happen at once.

In this case a number of other projects, that had had owner approval delayed, all got the go ahead just as this project was starting. In addition, a number of new projects 'came in the door', and hence there was a large demand for resources that the company did have.

The section heads did not carry out any formal resource smoothing across all the projects requiring resources. Resources seemed to be assigned on a perceived priority and on a 'whoever screamed the loudest' basis.

Scope

Scope is what is involved in undertaking the project, the extent of the work to be undertaken, what work is contained in the project (perhaps by defining what work is not included in the project, in effect the boundaries to the project) that leads to the end-product. Scope is fully described by listing all the activities.

There is much confusion over the usage of the term 'scope'. There is much loose usage of the term 'scope'. The biggest transgression is the sloppy, interchangeable use of the terms 'objective', 'constraint' and 'scope'; few people appreciate the distinction and why there is a distinction.

Commonly the carriage is put before the horse, such that the scope is said to determine the objectives and the constraints.

Getting to the scope is the process of working through the inverse problem associated with the 'means to the end-product'. Objectives and constraints are delineated, alternative work methods are proposed and evaluated against these objectives and constraints, and a best work method selected. This defines the extent of the work or scope.

As a project progresses, the objectives may change but more likely the constraints will change (for example, the work completed so far constrains what can be done next, assumptions on delivery time periods were inaccurate, . . .). This leads to scope changes.

The extent of the project work follows from a knowledge of the project objectives and constraints. Work established without reference to objectives or constraints could be expected to be non-optimal with respect to the objectives, and also possibly violate the constraints. Some people are able to rationalise this as being satisfactory, and some even rationalise it on the basis that they are being 'practical'. In reality, though, such people are unaware of how the scope follows from a knowledge of the project objectives and constraints.

Attention is paid to scope in order to indirectly [contain] deviations in cost, [schedule] and quality from that planned. Scope changes on projects are common and may be referred to as *variations* or *extras*. Some projects have a tendency for multiple changes leading to what is sometimes termed *scope creep*. Scope changes may come about through changed thinking, errors, changed values and circumstances, unforseen issues and so on.

People confuse a change in the end-product with a change in scope; the first will commonly lead to the second, but they are not the same. People also wrongly talk of control when they speak of 'change control'; invariably what is involved is a monitoring and recording of changes and perhaps some restrictions on changes. Changes then lead to replanning of the remaining part of the project.

Scope becomes better delineated as the project progresses. For many projects it is not possible to know the exact scope at the start of a project.

(Carmichael, 2004)

From objectives and constraints to scope

Scope derives from the project objectives and constraints. Figure 4.7 shows the steps involved. (Planning concerns itself with the 'means to end-product'

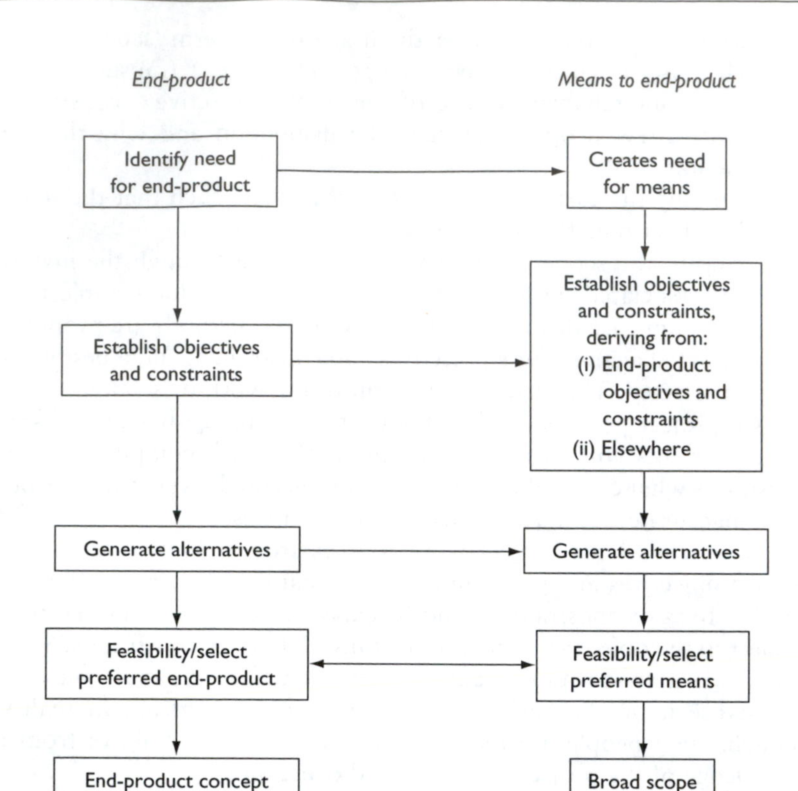

Figure 4.7 Broad design and planning steps.

problem. The parallel problem of design concerns itself with the 'end-product' problem.)

The means to the end-product is based on knowledge and experience relevant to that project's industry, and an understanding of that industry's projects. Ideally, all available ideas are evaluated. Viable alternatives/options are assessed including the carrying out of cost–duration trade-offs. Peculiarities of the project are considered.

Scope and planning

Scope establishment is part of the initial broad planning problem. In Figure 4.1, this is referring to the top left corner.

Extras feeding into and out of planning

Into

For planning to take place, supporting information is needed to feed into the basic problem-solving steps. Supporting information includes (Figure 4.8):

- Assembled and identified project data, for example, that obtained through a site inspection. Some familiarisation of the project environment and the nature of the work commonly takes place.
- Estimates of activity durations, associated resources (people and equipment), resource production rates and money to carry out these activities – availability of resources and money in time (when) and quantity (what). Estimates can be refined as the project progresses, and more information comes to hand; this in turn can lead to more accurate and more detailed plans.
- The behaviour of people (Chapter 7) including people's motivation, resistance to change and work ethos.
- Potential difficulties are anticipated. Forecasts are made and assumptions behind forecasts are tested.

Communication

The outcome of planning, together with the underlying planning assumptions, are communicated to the project stakeholders. Various forms of communication are possible, usually as documents (but can include physical and computer models) and verbal communication. The documents provide a basic reference and briefing for those who will ultimately be involved in the project execution. The documents allow project outsiders (including senior management) to review the plan, provide guidance and assist and, thereafter, support the plan throughout project execution.

If, following review, the plan is found to be satisfactory, the plan is implemented.

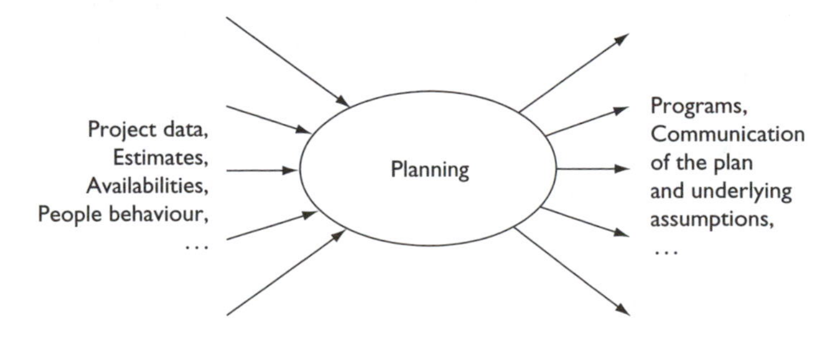

Figure 4.8 Extras feeding into and out of planning.

Table 4.1 Commonly used graphical forms of communication

Information with respect to	*Communication means*
Schedule	Bar chart, time-scaled network
Resources (people and equipment)	Resource plot, list; cumulative resource plot (S curve)
Money	Cumulative money plot (S curve), cash flow diagram, list
Production, work done; activity/project production rates; material usage	Cumulative production plot
Delays	Cumulative delay plot

There are a number of useful ways to assist the planner organise and present planning information. These take the form of charts, plots, activity date/time listings and so on. They might be regarded as the end-result of the planning process to date. Commonly used graphical forms of communication are indicated in Table 4.1 (refer Chapters 3 and 6 for descriptions of these).

As communication tools, there is a requirement for clarity, explicitness, being uncluttered, without superfluous information, and ease of understanding (intelligible) for all the end-users. Covering reports can be used to state assumptions, calculations and so on, and to keep the diagrams uncluttered. Colour can be used to improve interpretation.

According to Murphy's Law (extended): *The most important piece of information in any plan or document stands the greatest chance of being left out.*

Programs

General

Programs are a means of communicating the outcomes of planning. Ideally, some comment on the underlying planning assumptions accompanies the program, in order that end-users are well-informed, and can see the bigger picture. The forms programs take reflect what is trying to be communicated. Programs may be any of the following:

- (Connected) bar chart
- Time-scaled network diagram
- Cumulative production plot
- Activity date/time listings/timetable
- Marked-up drawings.

They ideally derive from a critical path analysis. They may commonly be presented:

- Over different project levels
- Over different time horizons.

Associated with such programs, there may be a need to also communicate resource and money requirements in the form of:

- Resource plots; S (resource) curves
- Budget, cash flow; S (money) curve.

and also possibly delay information.

End-users are acquainted with their individual responsibilities, possibly through responsibility matrices.

See Chapter 3 for the descriptions of these representations and terminology.

The program is the visible part of the planning process, used to communicate the results of planning. Programs are guides to assist the project work being carried out efficiently. They show the steps involved in the project work. Project participants can use the programs to see where they fit in the project. Durations and start dates might be arbitrarily imposed or decided in cooperation with all stakeholders, including contractors and equipment suppliers.

Programs give referenced (as-planned) baselines against which actual progress can be compared. Programs are continually updated, throughout a project, to reflect the latest information available. Sometimes bad practice is seen whereby a program becomes an end in itself, and is not used to assist replanning and project management.

For large projects, the number of activities may become unwieldy to handle together, and it may become necessary to create a number of convenient subprojects, or to group activities.

It is possibly too obvious to say that programs should be appropriate to the relevant end-user, and that they should be clear, user-friendly, at the appropriate degree of detail, and uncluttered in order to communicate the agreed plan. Some planners use the K.I.S.S. (keep it simple stupid) principle in order to improve understanding. Complicated programs are not recommended.

The maintenance of the program may be the responsibility of the contractor. Subcontractors may have a say in where they fit in the program, or may be advised when they are required to appear. Definiteness on these appearance dates assists subcontractors to manage their overall affairs. Vagueness can lead to a falling out between contractor and subcontractors as each may have a different view on what is expected of the subcontractors.

Programs over levels

On major projects, the communication of planning outcomes in forms such as programs could be expected to occur on several levels. The levels roughly align with organisational levels. Senior management may only be interested in broad issues. Middle management would show interest in more detail. Individual production groups would work with still more detailed information. There is no consensus as to the number of levels used or what each level is called. Figure 4.9 shows three levels, but less or more than this may be adopted depending on the situation. The higher the level, the lower the degree of detail. With reference to Figure 4.9:

- *Upper level* For senior management and the owner, a program is developed showing key work parcels and sequencing, milestones, and possibly contingencies (Figure 4.10).
- *Middle level* Middle management requires an indication of major work parcels and major events.
- *Lower level* Production management or workface management works off a program suitable for individual production groups.

Owners and senior management tend to be more concerned with overall results and not in the detail. In contrast, a foreman, say, needs detailed information over short periods to guide the workface employees. Intermediate detail may be required for contractual purposes when tendering, and may show major items and completion times.

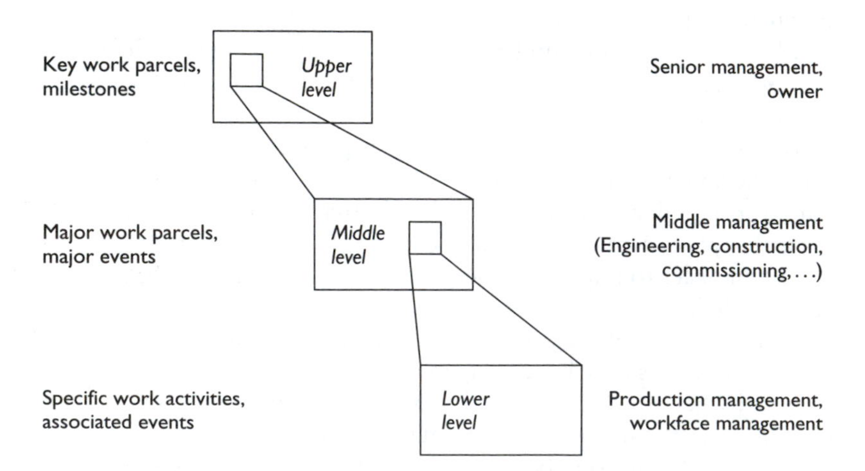

Figure 4.9 Example hierarchy of programs.

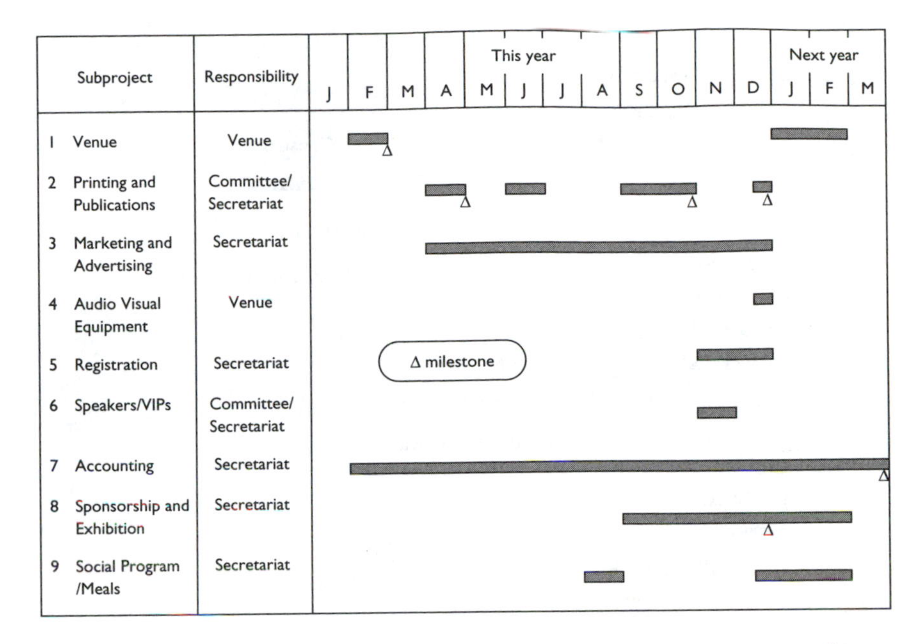

Figure 4.10 Example upper level program (simplified); example project involving a conference. (Work parcel names have been abbreviated because of space limitations – work items are implied.)

The programs at the higher levels are aggregations or integrations of programs at lower levels. The lower levels are equivalent to subprojects, sub-subprojects, . . . Resource, cost and duration information at the lower levels is folded up to higher levels (Figure 4.11). The better network analysis computer packages cater for different levels of reporting.

Program time horizons

For communicating the outcomes of planning, for any significant project, there is commonly a need for a number of programs covering different time horizons and associated different degrees of detail. There is no consensus as to the number of program types or the time horizon selected for each. Figure 4.12 shows four program types and a range of common time horizons, but different representations may be adopted depending on the situation. Figure 4.12 is based on the assumption that early in the project and throughout the project there will be a need for an *overall project program* (longest time horizon program) showing major work parcels, milestones and key resources.

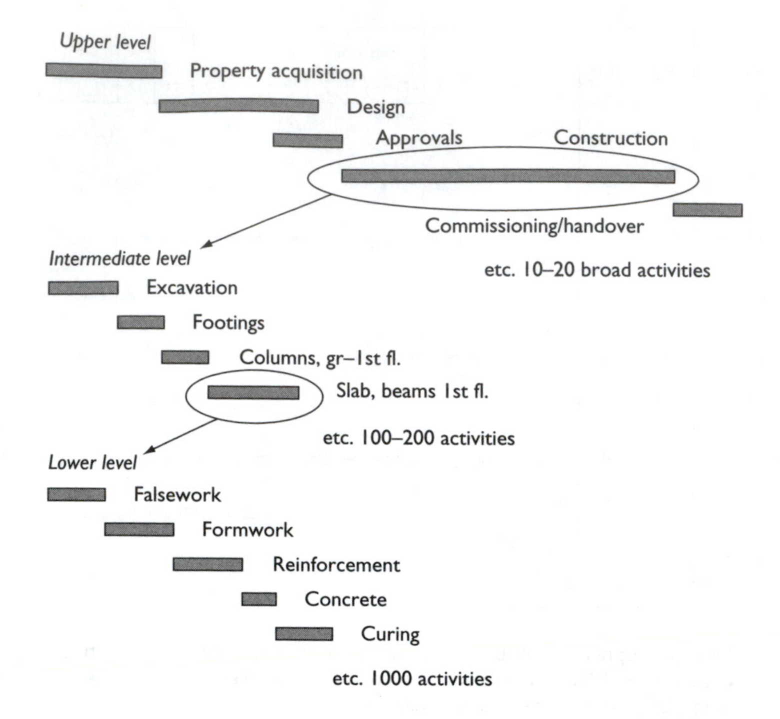

Figure 4.11 Different levels of programs; example, construction of a multistorey building. (Work parcel/activity names have been abbreviated because of space limitations – work items are implied.)

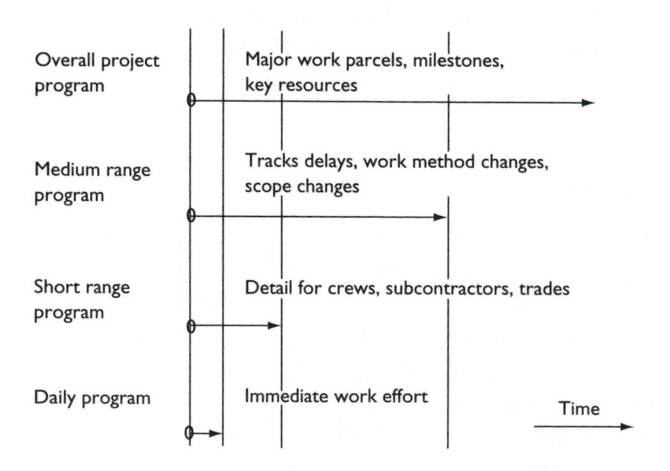

Figure 4.12 Example programs with different time horizons.

As the project progresses there will be a need for a *medium-range program* showing detail over the next, say, 2–3 months and updated, say, monthly. Updates will take account of, for example, delays, work method changes and scope changes.

At the workface level for individual crews, subcontractors and tradespeople there will be a need for a *short range program* showing detail over the next, say, 2–3 weeks and updated, say, weekly.

There is also a place for a *daily program* (shortest time horizon program) showing immediate work effort. This program need be no more than a list of items.

Where the project is being fast-tracked, and to ensure that the project proceeds speedily, a *starter program* may be issued along with the preliminary project program. The starter program lists the important activities to be performed during the first, say, 30–90 days of the project while other programming is being developed.

Detailed execution (fabrication, construction, . . .) programs are developed during the design phase prior to the start of any physical work.

The above time horizons are examples only. For any given project, the time horizons are selected to suit.

Summary

Table 4.2 attempts to summarise the various types of programs used on projects. The levels correspond roughly with the levels in an organisational structure. There is no consensus as to the usage of terminology in Table 4.2, or to the number of rows in Table 4.2.

Some organisations go to the trouble of creating a summary document, describing all programs, when they are to be produced during project execution, what they must contain, and to whom they will be directed.

Table 4.2 Example summary program forms

Level	*Coverage*	*Extent or scale*	*Unit*	*Degree of detail*	*End-user*
Upper	Outline	Entire project	Month	Low	Management
	Outline/broad	Project design; Execution/ implementation	Week	Low/medium	
Middle	Execution/ implementation	Execution/ implementation period	Week	Medium	Project office
	Short term	5–10 weeks	Day	Medium/high	
Lower	Weekly	1–2 weeks	Half-day	High	Workface
	Daily	1–2 days	Hour	High	

Case example – Aerospace industry

This case example discusses the use of schedules and network diagrams used within an aerospace company for the management and reporting of progress on its projects.

The master schedules formed part of a contract with the respective customer, monitored by both parties at formal review meetings with actions required if the project falls behind schedule.

Resource planning and cost impact were not performed adequately until the use of PC-based planning packages. Formerly, schedules were shown as both broad hand-drawn bar charts and cumulative production plots (linear schedules). The packages have enabled more detail to be included in the planning.

The master schedules tended to only show major events, or milestones, and total duration. For example, the manufacture of 200 metal components for use in final assembly was shown as one bar, not 200 bars, with a total duration. Possible inclusions within the schedule included:

- Despatch date
- Paint start and finish
- Final assembly
- Carbon fibre component start and finish
- Metal component manufacture – in batches
- Raw material requirements, usually shown as a milestone.

Cumulative production plots were also used in broad form (Figure 4.13).

Because each completed component was expensive, the above inclusions were repeated for every product despatched over a two- to three-year period.

A broad network diagram, hand drawn, was put together before the first master schedule was completed. The activity durations, and the logic of the network diagram, were reviewed continually to reflect the true manufacturing situation.

During the implementation phase, the (broad) network diagram and (broad) master schedule were changed to reflect the new situation (ahead/behind schedule) and when the company would be back on schedule.

The introduction of PC-based scheduling packages bought about changes (more detail) in the presentation of master schedules.

The company's planning practices included liaising with the customer in establishing milestones, liaising with production as to realistic durations, and monitoring actual progress against as-planned.

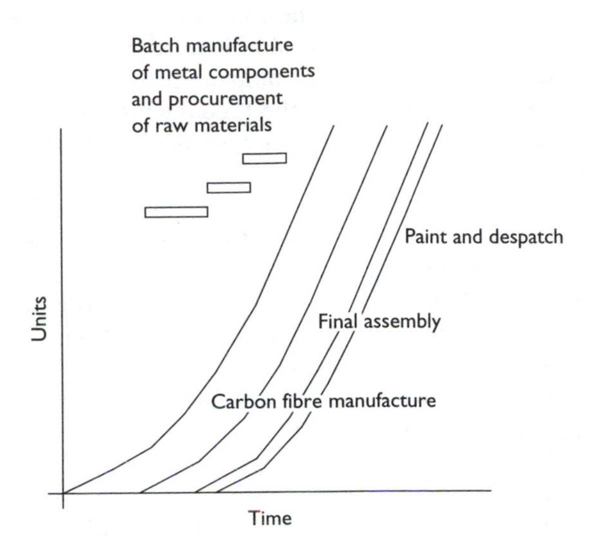

Figure 4.13 Example cumulative production plot. Units are the quantity produced during the contract period or part thereof.

Formerly, without the critical path being known, the people responsible for performing the actual work were not aware of the importance in adhering to designated schedules. This also created a problem for the project manager.

Where progress fell behind schedule, recovery schedules were derived to bring the overall project back into line. Under these circumstances, each aspect of the network diagram was reviewed, relevant people were asked to reassess the activities and when they could be started/completed. This would lead, amongst other things, to overlapping activities.

Case example – Water treatment plant

The company in this case example is a designer, manufacturer, supplier, installer and operator of engineered solutions and equipment for the water and wastewater industry. The company has projects of various sizes and complexity and in various stages of completion underway at any one time, and spread geographically. This case example focuses on a project to develop a water treatment plant for a mine.

The company was awarded a contract by the mine for the design, supply, installation and commissioning of a water treatment plant to supply demineralised quality water. The contract required a completion date of one year; there were no separable portions.

As part of its tender submission the company provided a program in simple bar chart format setting out the fundamental elements and work order of the project. This program contained some thirty elements without background assumptions.

Immediately upon contract award, the company provided a very detailed program. A work breakdown structure gave the fundamental elements of the work. The program consisted of some 400 activities, and identified the critical path. An S curve (money) was also included.

Two months after starting work, by negotiated variation, the mine increased the output of the plant. This required a revised process and alternative equipment to provide for the increased flows. As a result of the process change, the project was split and two separable portions were formed with staggered handover dates.

A revised program detailing the new process and methodology was submitted, this time with some 500 activities. Again the critical path was identified and an S curve (money) given.

Under the terms of the contract, a detailed monthly progress report was to be submitted as an ongoing task. Each report contained an updated program with progress reported against activities and the S curve, together with derivative programs such as four-week look ahead, engineering and procurement detail programs.

As the project unfolded, many special in-house subprograms were produced as working tools. Procurement programs for items such as plant valves were developed. The procurement of the project valves as a total group was identified on the subprogram as engineer, size, enquiry, select, order, manufacture and delivery work without regard or distinction between the types, makes and manufacturers. The subprogram catered for all the variables and delivery of the selected valves to the site, as and when required.

Exercises

1. Programs are usually prepared with different degrees of detail for different end-users. For example top management requires broad overview information, while people at the workface need very detailed task-by-task information.

Consider a project with which you have been involved and consider the number of levels of people involved. How many programs would be necessary and what degree of detail would go into each program?

2. Rank, in your opinion, the following poor practices in programs:

- Too much or too little information
- Development without foresight as to the program end-users*
- Lack of continuous updating
- The use of unsuitable computer packages
- Lack of identifying programs as part of contract administration procedures
- Lack of support of programs with planning information (assumptions on methods, resources, constraints, . . .)
- Completion times given based on inadequate data.

[* Program end-users refer to those who are to use the plan, the so-called 'doers'.]

Add others to this list, that you may be aware of, and also rank all the practices that, in your opinion, cause the greatest detriment through to the least detriment.

What are the causes of these practices? A cause–effect diagram might be helpful in answering this question.

What suggestions would you make to eliminate the practices?

3. For a project with which you are familiar, what would be the most important tasks that should be incorporated in a preliminary project program/schedule?

4. Consider a project with which you have been involved and consider the time horizons of the programs involved. How many programs were necessary and what time horizons were used in each program?

Chapter 5

Planning over levels

Introduction

Simplifications

Planning is simplified in a number of pragmatic ways, one of which is over levels (Figure 5.1):

- The project (work) is broken down (hierarchically) to lower level subprojects, sub-subprojects and so on to activities, elements and constituents. Subprojects may be chosen such that the interaction between subprojects is minimal and is commonly neglected, permitting subprojects to be planned independently of other subprojects. The same holds true where multiple projects are combined on an organisation level. However, interaction at levels lower than subproject is commonly strong.

  Interaction occurs because of the work method (implying an order/sequencing of work). Iterations (between levels) can largely be removed by the planner being aware of what is happening on other levels when solving an individual level problem. The levels are uncoupled as far as possible, for tractability reasons, to make the level problems less complicated.

Planning goes from the general to the particular, from coarse to fine (broad to detailed) as the levels descend. Detail emerges from lower levels.

Work breakdown

The work in a project may be systematically broken down to subprojects to sub-subprojects and so on to component activities. Such a work breakdown can be represented as a tree-type diagram called a *work breakdown structure* (WBS). The WBS, however, says nothing of interaction (or interrelationship) between subprojects, between sub-subprojects, . . . , between

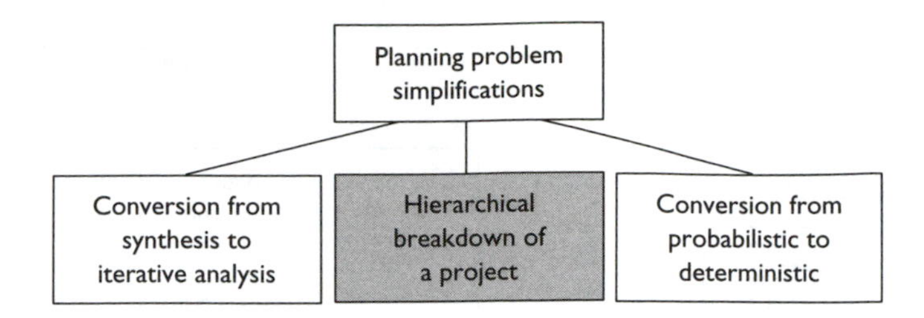

Figure 5.1 Simplifications to make the planning problem more tractable.

activities. A work breakdown also says nothing about events, though loose practice by many writers confuses this.

A systematic breakdown is preferred over an *ad hoc* approach because of the increased likelihood of identifying all activities.

Hierarchical modelling

Levels

A model of a project for planning purposes can be built from the bottom-up using component parts. Figure 5.2 shows two possible representations. Constituents feed up to elements; elements combine to give an activity; activities combine to give a project (and projects can be combined to give a 'program'). In Figure 5.2, subproject- and sub-subproject-type levels (equivalently work packages) can be inserted, but add nothing new to the following development; a sub-subproject and a subproject (and work packages) are no different in character to a project. Similarly, inserting a 'program' level adds nothing new; a program is no different in character to a project.

Formalism

Chapter 9 develops formal project models.

Control

The following examples of controls or actions are not totally exhaustive. Controls chosen (actions that are taken by the planner) depend on the planner's experience and expertise, as well as the situation and a bit of creativity. Some choices may be better than others, but generally planners are after a satisfactory solution rather than some theoretically optimum solution (within an iterative analysis approach to planning).

Controls at any level contain lower level control information.

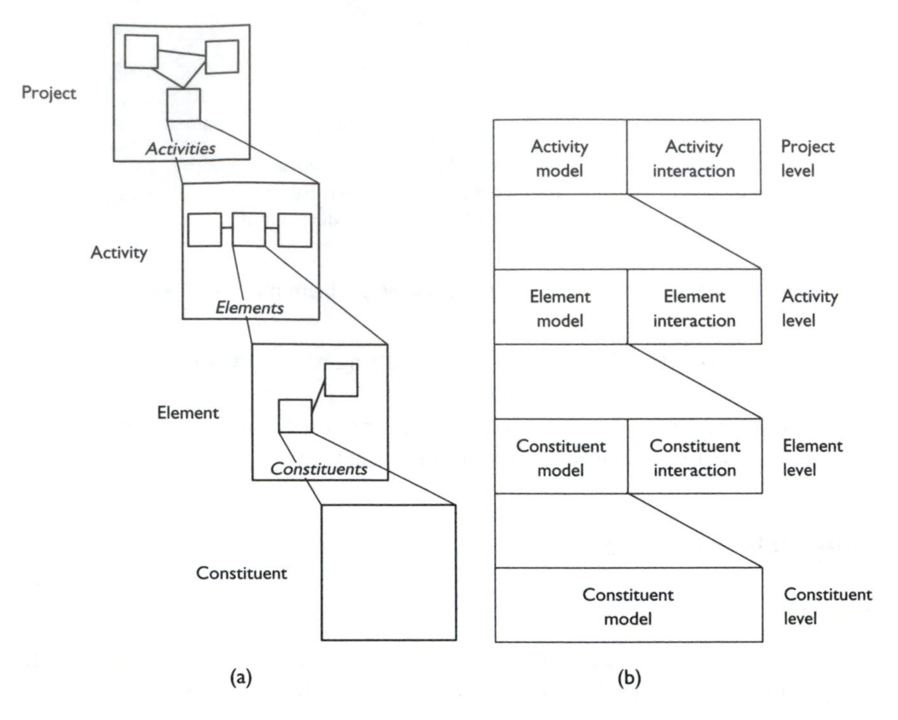

Figure 5.2 Building up a project model.

Project level controls

Controls at the project level relate to the work method perhaps involving the interaction of activities (and this implies an order/sequence of work). Examples include:

- Examining alternative methods of work. This will lead to alternative network logic. Examples include outsourcing or contracting out some work instead of doing the work in-house, different techniques, different ways of conducting activities, different work sequences or logical ordering of activities, different managerial and technical approaches to the work, using machinery instead of labour, and prefabrication rather than an *in situ* work method.
- Using overlapping relationships between activities. In place of more usual finish-to-start (F/S) relationships, start-to-start (S/S), finish-to-finish (F/F), or start-to-finish (S/F) relationships might be possible, as well as finish-to-start (F/S) relationships with lead time periods.
- Run concurrent/parallel activities. Some activities can be carried out sequentially, some simultaneously.

- Expediting materials; introducing 'just-in-time' (JIT) or other inventory practices.
- Extending the project duration. All activities then become non-critical.

The simpler planning problem fixes the work method and only leaves resource usage and resource production rates as the variables to be determined. That is, the controls at the project level are fixed, and only lower level controls remain to be selected.

Activity level controls

At the activity level, control relates to resource usage within an activity:

- Total quantity – Use additional/less resource numbers, for each resource type.
- Distribution in time – Compress activities, lengthen activities and split activities.

In conventional planning parlance, this is achieved through, for example:

- Adjustment of activity resource numbers (each resource type) – distribution over the activity, decreased quantity or increased quantity. Different ways of using resources (or money) on activities. This may go under the name of resource redeployment or redistribution. Adjusting resources may lead to changed activity durations.
- Compression of activities. Compressing the durations of activities involves shortening the activity durations; this usually comes with an additional and a more concentrated resource requirement (and an increased cost).
- Lengthening activities, though this may come with lower production and a cost penalty, sometimes called uneconomical drag out.
- Splitting activities that can be split. Splitting implies starting an activity, stopping the activity, starting the activity after a break and so on.
- Making use of any activity float. Typically the start dates of non-critical activities are moved. Note that shifting the start dates of critical activities will extend the project completion date.
- Changing the work calendars.

Element level controls

At the element level, there is a collection of resources (different resources and possibly different numbers of each resource type). Control relates to the selection of the number (quantity) and type of each resource. For example, working multiple shifts or overtime fits within this meaning. Different numbers and types of resources lead to different production.

Resource information can be collectively measured in a money unit.

Constituent level controls

The constituent level is the basic behaviour level of a resource (person, or piece of equipment). As such, the controls are the resource production rates or equivalents of each resource type. Resource production rates, for example for people, can be altered by the use of a carrot (incentive, reward) or stick, changing the work conditions, having the correct tools and specialisation. Resource production rates are different for different pieces of equipment.

Work breakdown

General

Work breakdown is a top-down approach, as opposed to the hierarchical modelling, given in the previous sections and as discussed later, which is developed from the bottom-up.

A project (scope) can be decomposed into subprojects and then into sub-subprojects and so on down to the activity level (but usually not finer). The lower levels represent finer detail. The process of decomposing a project might be called *scope delineation* (or scope definition) by some writers. The term 'work breakdown structure' (WBS) is used to refer to the resulting systematic decomposition (as opposed to an unstructured activity list where there is a possibility that some activities have been overlooked).

What is not evident from the work breakdown structure is the degree of interrelationship between its components, or the reliance of some on others, and how the work breakdown structure has been used to separate individual components.

Two wrong and confusing practices that can be seen with some writers are:

- The inclusion of events (and end-products and 'deliverables') in a work breakdown. A work breakdown says nothing about events. An event is the start or end of a work item; it is not work. The systematic nature of a work breakdown is destroyed by the *ad hoc* introduction of events and deliverables. This is not to say that events and deliverables are not important for project management purposes, but rather events and deliverables have no place in a work breakdown; as well arbitrary or *ad hoc* practices are to be avoided.
- The labelling of entries in the work breakdown as non-work items. For example, a work breakdown entry may be labelled as 'book' when it should be 'some work involving the book'.

The work breakdown structure provides a common framework for dealing with all aspects of planning a project (in an iterative analysis sense) – a divide-and-conquer approach. It enables a clear understanding of the scope, and reduces the likelihood that activities will go unthought of (thereby avoiding disruption to the project rhythm, the incurring of extra duration and resources and hence extra cost, and the consequent effects on team morale).

Representation

Work breakdown may be represented as a *tree* diagram (Figure 5.3), or appropriately indented dot points, appropriately numbered items (Figure 5.4) that recognise the hierarchy involved, or as in Figure 5.5.

With respect to Figure 5.4, there is no set standard for numbering schemes, or numbering for accounting purposes. Rather, the schemes may reflect different project characteristics and different organisational ways of dealing, for example, with money.

Example

Consider a project involving the staging of a conference. Major subprojects would centre on work associated with:

- Venue
- Printing and publications
- Marketing and advertising
- Audio-visual equipment

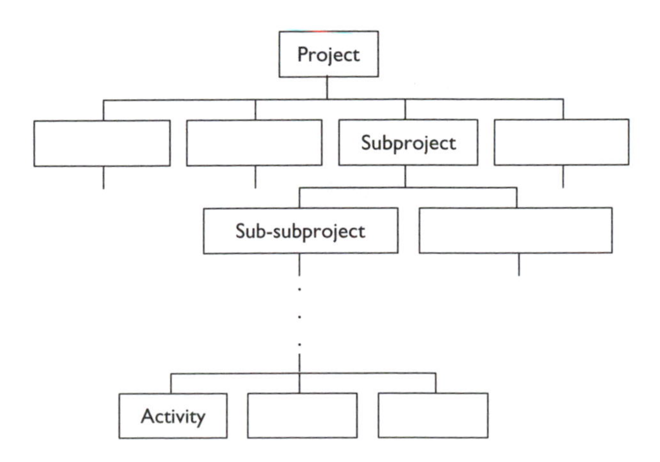

Figure 5.3 Work breakdown structure (tree diagram form).

Project
1000 Subproject
1100 Sub-subproject
1101 Activity
1102 Activity
1103 etc.
1200 Sub-subproject
1201 Activity
1202 etc.
1300 etc.
2000 Subproject

3000 etc.

Figure 5.4 Example hierarchical breakdown and numbering of activities.

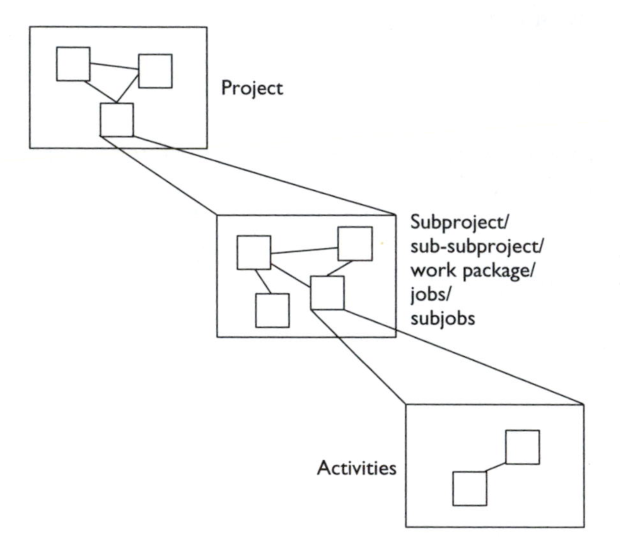

Figure 5.5 Hierarchical decomposition of a project.

- Registration
- Speakers/guests
- Accounting
- Sponsorship and exhibition
- Social program/meals.

These may be further subdivided into their component parts. Figure 5.6 shows how this may be organised hierarchically in a tree diagram. The tree

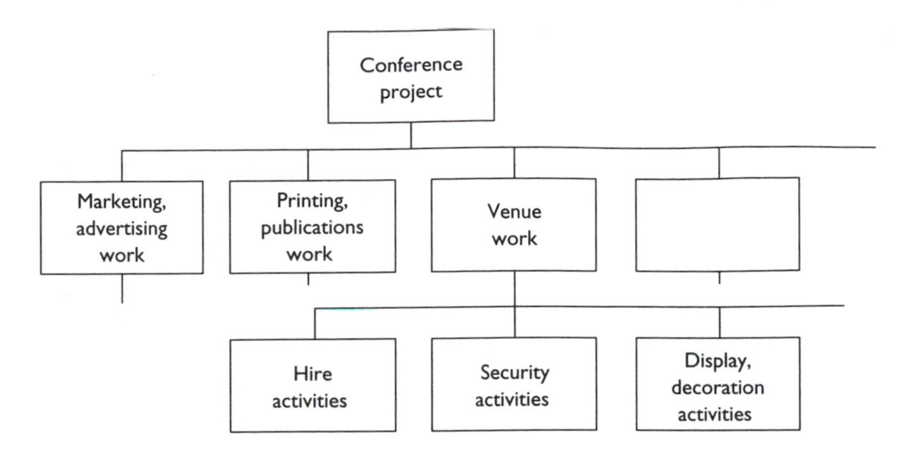

Figure 5.6 Tree diagram conference example.

diagram (Figure 5.6) says nothing about the order of and the interaction among the work.

Subproject choice

Subprojects may be chosen at the convenience of the planner but might follow a breakdown according to:

- Phase (chronological); time period phasing
- Work type or nature; resource groupings
- Work parcels; the way the work is to be carried out
- Contracts
- Region, location or geographical area
- Organisational breakdown
- (Existing) cost codes, cost centres, cost headings; project cost and data summary reports
- The nature of the planning network; activity groupings
- Function.

The particular subdivisions chosen in the breakdown will reflect some convenient way of parcelling and doing the work. Different organisations and people, if given the same project, may choose to subdivide it differently.

It can be helpful to list the subprojects in the order in which they might normally be expected to occur on the project, but this is not necessary.

Irrespective of the breakdown adopted, the collective bottom level of activities should be the same. All the work that has to be done to complete a project is contained within the work breakdown structure.

Commercially available computer packages used to assist planning commonly adopt a work breakdown structure as a basis for data input.

Lower levels can be integrated or aggregated (folded up) to higher levels for reporting or communication purposes. Any level is the sum of the immediately lower level work.

With this aggregation, comes an aggregation of duration, resource and cost information, such that information on durations, resources and cost can be given at each level. Monitoring (project production/progress and resource usage), reporting (planned and actual production/progress and resource usage comparison) and replanning can be carried out at each level.

Number of levels

The number of levels needed to describe all the project work might range from about two to about five. The *lowest level* is chosen to give manageable size entities or identifiable pieces of work; the activities are able to be characterised adequately in terms of duration, resource and cost estimates; responsibility and authority can be assigned to each. Until the breakdown reaches the level of activities, it is difficult to undertake estimating accurately. Resourcing, costing, scheduling, uncertainties, progressive measurement and reporting relate to these activities and estimates. What is 'adequate' for estimating purposes may change over the course of a project. The breakdown may evolve as the project progresses, as the project becomes more definite.

Different stakeholders may choose to decompose to different levels. For example, a project manager may stop at the level of trade contract packages, while each trade contractor will subdivide its work package to smaller items.

Complete decomposition may not occur in some instances. While certain areas may be well defined and broken down into clear and measurable components, others may be defined only to a level where a general understanding of the activities is evident. This may be due to the complicated nature of these components or further breakdown may be considered unnecessary. This lack of complete breakdown may lead to some problems later in the project with some planning and design information being found to be incomplete.

In practice, the development of a work breakdown structure may proceed simultaneously from the top-down and from the bottom-up.

At the lowest level of this hierarchical breakdown of a project the entities are manageable components, namely individual activities. These are of relatively short time span. At intermediate levels the entities are of longer time span.

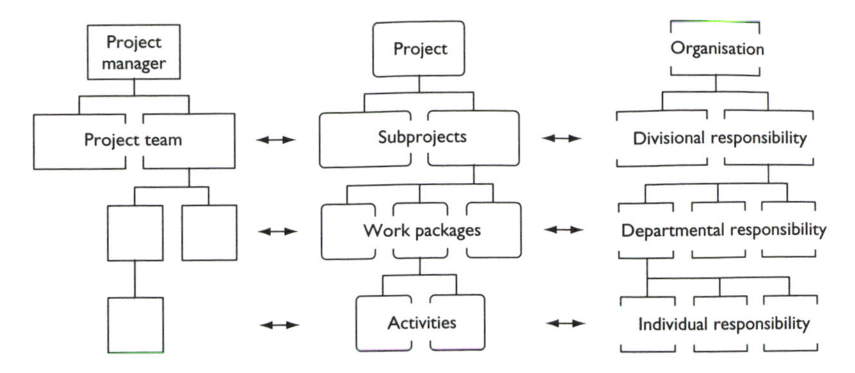

Figure 5.7 Possible relationship between work breakdown and organisational breakdown.

Einstein is said to have remarked that: *Everything should be made as simple as possible, but not simpler.*

WBS relationship to OBS

The project breakdown may bear a close relationship with organisational breakdown (Figure 5.7) – project team structure when applied to a project, or *organisational breakdown structure (OBS)* when applied to an organisation and its functional lines – a representation of the personnel involved and their interrelationships.

Change

Change may be viewed as:

- Changes to the nature of activities within the work breakdown structure and
- Required additional/less activities.

The second category represents scope changes. Changes to the nature of activities within the work breakdown structure may not normally be regarded as scope changes, though from a contractual viewpoint might be regarded as a variation.

Example – Writing a book

Consider the writing of a book involving the researching of numerous sources, and contributions by different authors. A breakdown of the work

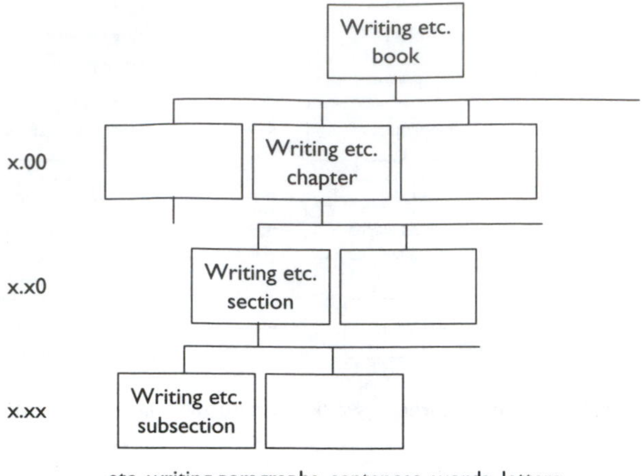

Figure 5.8 Example WBS for compiling a book.

might be viewed according to Figure 5.8. Lower subdivisions to writing paragraphs, sentences, words and letters could be considered but are perhaps too small to be manageable.

The development of the WBS proceeds from the top-down but relies on the end-users (writers) providing input of what is required at the lower levels.

Cost account numbers could parallel the chapter/section/subsection numbering scheme. A responsibility matrix could be set up identifying each chapter/section/subsection with a writer, checker and editor. Duration, resource and cost estimates for the writing of each chapter/section/subsection can be made, aggregated to give an overall duration, resource and cost estimate for the book, and a schedule and budget for its writing. Monitoring and replanning the writing of the book can be directly related to chapters/sections/subsections.

Example – Building a house

Consider the building of a low-cost house. The component activities might be grouped as in Figure 5.9.

Figure 5.10 shows schematically this breakdown.

Example – Road project

A breakdown for a road project might look something like Figure 5.11.

- 1 Preparatory/preliminary work
  - 1.1 Erection of site facilities
  - 1.2 Survey and layout site
- 2 Foundation work
  - 2.1 Excavation
  - 2.2 Setting formwork and reinforcing steel
  - 2.3 Concreting
- 3 Floor construction
  - 3.1 Setting formwork and reinforcing steel
  - 3.2 Concreting
- 4 Wall construction
  - 4.1 Bricklaying
  - 4.2 Installation of door and window frames
  - 4.3 Plastering
  - 4.4 Painting
- 5 Roof/ceiling construction
  - 5.1 Roof framing
  - 5.2 Placing roof tiles
  - 5.3 Installing ceiling
- 6 Plumbing/electrical work
  - 6.1 Excavation, install septic tank
  - 6.2 Laying drainage, sewerage pipes
  - 6.3 Laying water pipes
  - 6.4 Running electric wiring
- 7 Other work
  - 7.1 Laying driveway, footpath
  - 7.2 Gardening
  - 7.3 Cleaning up

Figure 5.9 Example work breakdown for the building of a low-cost house.

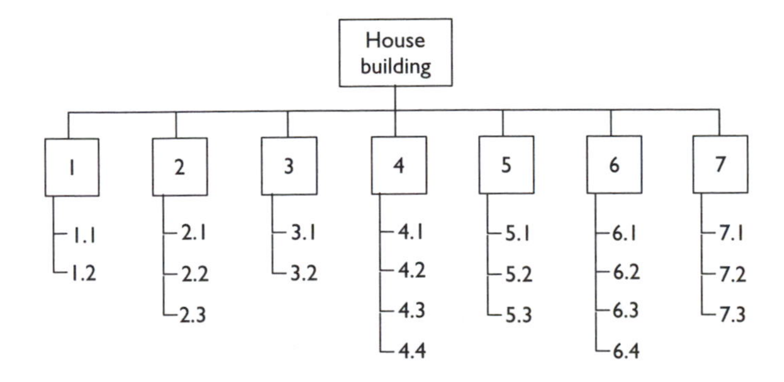

Figure 5.10 Example work breakdown for the building of a low-cost house.

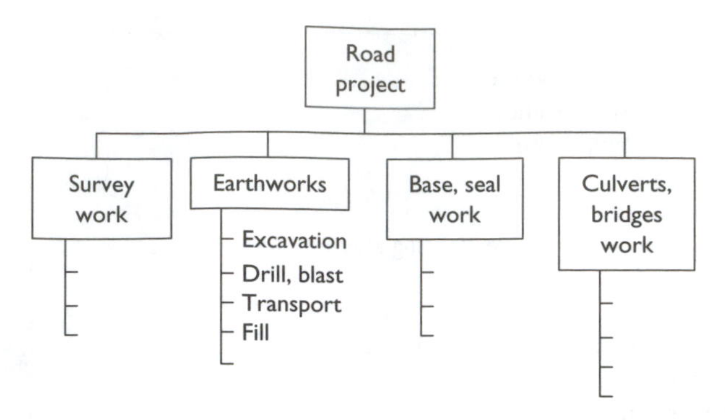

Figure 5.11 WBS for an example road project.

Case example – Truck research

This case example describes the work breakdown structure for a project involving the examination of the operation of mining haul trucks, from the viewpoint of the owner of the mine and the trucks. Research work is implied, but not stated, in the breakdown that follows.

The case example relates to a research and development project established due to inadequate brake, transmission and differential life of mining haul trucks. The overall aim from the owner's view was to reduce costs associated with the operation of the haul trucks. The mine looking after these trucks had to take the research results and address them wherever practical. After preliminary research into the trucks, it was noticed that the majority of costs incurred were in maintenance, and in particular relating to brakes, transmission and differential life. The mine was uncertain as to why these trends were occurring and hence was required to do further research. A research and development project was established.

Due to the expansive nature of the project, it was first divided into research work associated with three subprojects:

1. Issues relating to brake life
2. Issues relating to transmission life
3. Issues relating to differential life.

Procedures used and decisions made, up to and including this (first) division, all constituted initial stages of planning. A problem was identified and a project developed to investigate that problem. There were many issues which affected the costs of haul trucks; honing in on maintenance issues already created a subproject, because there were other ways that costs on the haul trucks could be addressed. The mine had effectively narrowed down an area based on collected data, where they thought that significant cost savings could be produced. However, if this path was proved to be wrong, then they needed to track back to this first division and follow a different path.

Once the overall project had been given a definite subproject to follow, this was further broken down into three distinct areas based on truck components. At this point, creating subprojects helped to keep the overall project's aims in sight whilst breaking the project into subprojects and sub-subprojects enabled easier management and planning. If work breakdowns had not occurred, through a systematic approach backed up by previously collated data, then identifying an area where most expense on the trucks was occurring would have been a burdensome task. It also followed, that had the first division not been backed up by history files, experience etc., honing in on the correct components that were of the greatest cost would not have been possible. Not much effort was spent looking into the detail of why or how the problems were occurring, but rather what was costing. The 'why' details came forth later in the project breakdown.

Narrowing in on the issues relating to brake life, a new set of aims (sub-aims) was defined, to assist in keeping the overall project's aims on track. A sub-aim of this project was to identify the limiting factors of brake life for the haul trucks. Once this sub-aim was satisfied, limiting factors and therefore areas with associated cost were to be identified and were to help contribute to answering the overall aim.

Even though the subproject had been divided into sub-subprojects to have reached the above stage, it was still too cumbersome to allow for factual and clear communication of identified problem areas; so it was once again broken down smaller, that is to sub-sub-subprojects. There were many factors which may lead to limiting the brake life of these trucks at this level. Two clearly separate issues relating to research on brake life were identified (second division):

1.1 Factors relating to the application of the trucks
1.2 Factors relating to the design of the trucks.

Within each of these, a further breakdown was evident; this next breakdown addressed research work areas that were categorised according to the above factors (third division):

1.1.1 Speed in relation to brake effectiveness
1.1.2 Haul road profiles and effect on braking system
1.1.3 Loading of vehicles

1.1.4 Operator considerations/human issues
1.1.5 Maintenance considerations.

and

1.2.1 Characteristics of fluid in hydraulic system
1.2.2 Material of brake pads
1.2.3 Design characteristic monitoring
1.2.4 Different types of braking systems
1.2.5 Environmental effects
1.2.6 Pressure system and fluid flow for hydraulic brakes.

The second division was made by the subproject manager so that the sub-aims could be clearly met. This division was based on brainstorming all possible contributing factors to brake life. After brainstorming it became evident that two main factors could be contributing to the brake costs. The items which fell into these two main categories went together to make the third division. The procedure followed to come to the second and third divisions differed from that followed to come to the first. This was because these later two divisions were the initial focus on not only what could be contributing to the brake life but also why these categories could be contributing to the brake life problems. The second division and the ones that followed were based on hypothetical possibilities; there was no firm previous data that could have been collected to give the best direction.

Activities and work procedures were then identified and were combined in addressing the set of aims associated with the third division. These were seen as one of the last steps in the breakdown tree. The work procedures highlighted how the third division aims contributed to the overall project's aims. Future steps then became specific activities rather than areas and so on.

Case example – Bridge construction

This case example describes the work breakdown structure (WBS) for the construction of a bridge from the viewpoint of the contractor.

For the incrementally launched box-girder bridge, the breakdown chosen stemmed from a combination of planning requirements and the ease by which the project manager could monitor, manage and record construction activities and costs associated with each subproject. General agreement and past experience between the planning and construction teams limited the number of work breakdown levels. The work breakdown structure for the project consisted of four main subprojects, namely work associated with:

1 Substructure
2 Piers and abutments
3 Box girder
4 Completion of superstructure.

Each of these subprojects was further divided into work associated with:

- Plant and equipment
  - Hire
  - Maintenance
- Concrete
- Reinforcement
  - Subcontract
  - Materials
- Formwork
  - Subcontract
  - Materials.

Separate work breakdown structures existed for activities associated with site mobilisation and demobilisation, electricity, tools, safety, consumables, site overheads and so on.

The breakdown of the project into these four major subprojects was intended both for ease of record keeping and for usefulness of the records kept. Whilst box-girder bridges may have different piling (substructure) and pier column arrangements, information on the costs of building that particular type of box girder, regardless of the foundations, is useful and hence was recorded. Similarly, by separating the substructure work, records were now available on the costs and production of the barge, the piling hammer and the manufacture of pre-stressed concrete piles. This enabled bridge records to be used for a variety of other projects, for example wharves.

The work breakdown structure also followed the course of the job. By relating the work breakdown structure to the master plan, the project manager was able to monitor the project, in both schedule and cost senses. Schedule management was achieved by being able to systematically close-out subprojects. Cost savings in a completed subproject could be used to offset the increased costs in another subproject, enabling the project to remain within overall budget.

The subsequent breakdown into work associated with plant and equipment, materials and labour was largely a breakdown based on costs. The majority of plant and equipment for the project was obtained on internal hire; thus the breakdown assisted in the preparation of monthly accounts. Materials were assumed to be estimated accurately at tender stage, and subsequent blow-outs in this area pointed to estimating problems. Labour costs, the greatest variable, gave an indication of the management of the project.

Case example – Computer system upgrade

In this case example project, the work breakdown structure (WBS) was not clearly understood by the stakeholders, resulting in confusion, frustration and near disaster.

The project, with its end result of upgraded technology throughout a firm, involved:

- Upgrading all hardware
- Upgrading all hubs and associated cabling
- Upgrading all operating systems
- Updating application software
- Training of all staff in the new software applications.

The first three of these were achieved with only relatively minor troubles and concerns, while the fourth proved to be a major headache due to the number of subprojects it contained and the necessity to outsource five of these subprojects to five separate organisations – a decision based on a lack of required expertise and experience within the firm.

At the completion of planning, and by the time it came to update the application software, it was assumed by all stakeholders that 'all bases had been covered', the stakeholders understood their roles and those of others within the project, and nothing had been overlooked. Various companies were working together on the document macro and template developments and design of a conversion filter for existing documents. Other companies were working on the design of the document management system profiles for the new documents to be created and their conversion. Everyone, it seemed, was actively engaged in working towards converting existing documents and document management system profiles over to their new format, and preparing for the creation of new documents and profiles.

However, no one had been assigned responsibility for the actual conversion of the existing word-processing documents. Each stakeholder assumed that another was responsible for this activity because everyone had seen and worked with the conversion filter at some stage and had discussed it at some point during the project. It was assumed by all, therefore, that preparations for the commencement of training were in hand.

Fortunately, this error was discovered the day before training was to commence, soon enough to avert a near disaster, should it have remained undetected. Training was to be in groups of six staff members at a time, offsite, in all aspects of the new system. While each group was in training, several other activities were to occur concurrently: the rolling out of the new desktops onto their PCs, the conversion of their word-processing documents and their document management profiles.

Had their documents not been converted upon their return from training, they would not have been able to access critical work. Their document profiles would have been converted to the new document management system and their PCs would have been ready with the new system but they would not have been able to find any existing work. It would have been impossible to convert their documents across in time for them to work for the first few days after training, and because thousands of these documents were vital legal documents, this was a situation which would have had serious repercussions. It could have been easily avoided by better planning and closer attention to detail.

The project had continual changes within each subproject, as the software was developed.

Subproject work that was done by outside contract lacked a scope statement defining the extent of the work. A few faxes confirming verbal quotations were all that were present, but even these did nothing to state the overall area of responsibility or scope (of work) to be undertaken by each contractor or the firm.

The project manager was appointed from within the firm. Despite having no prior project management experience, he had the necessary skills to understand the scope (of the project), and was a good manager, with good people and communication skills. He also had the support of senior management, who nevertheless expected him to add it to his normal duties. As a result of this, the person who was expected to be the main impetus behind the success of the project was so busy and so much in demand that, instead of being able to dedicate a worthwhile period of time and attention to ensuring that each activity was performed, was constantly spreading himself around in smaller, superficial time lots, without being able to really manage the project. When he requested assistance from senior management it was refused, with the reason that 'no more resources (people) could be spared'. As a result of this limit on human resource numbers, other problems arose. One of these was that documentation of the progress of the project was sparse. Perhaps if such progress had been documented, it would then have drawn attention to the oversight on the activity of converting the documents.

Documentation could, at a minimum, have included a detailed WBS, with activities delegated to appropriate members of the team. It may have drawn attention to the impending problem.

Had the documents not been converted in time, the inevitable inability to access work would have occurred for the trainees. This may have had the effect of causing frustration, accusations and, possibly, arguments. It also would have caused a blow-out in the schedule and budget because the firm possibly would have had to then insist that some of the trainees get involved in the conversion process, thereby taking them away from other work.

Iterative analysis approach

Generally, planning over levels is practised in conjunction with an iterative analysis mode of attack, primarily because the synthesis version is too difficult (Figure 5.12).

The iterative analysis approach follows Figure 5.13, showing the feedback loops, and in particular the feedback loops between levels. Figure 5.13 is saying the same thing as Figure 2.9.

The initial solution guess is commonly based on past projects, experience and knowledge of the industry.

The analysis and evaluation will generally involve network analysis and the use of time-scaled networks, bar charts, cumulative production plots and so on. The analysis may be assisted by 2D and 3D computer visualisation, scaled models and simulation.

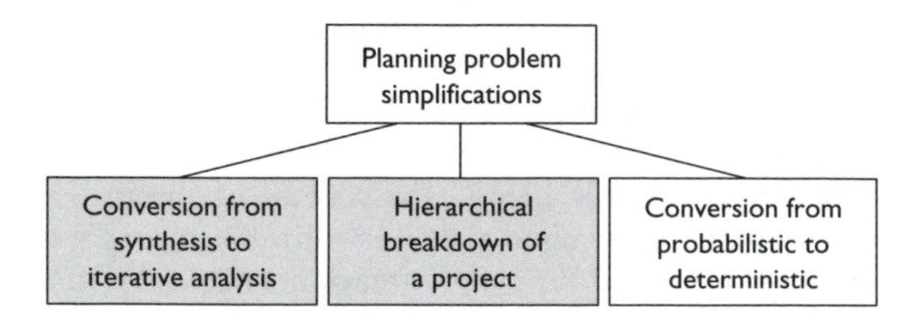

Figure 5.12 Simplifications to make the planning problem more tractable.

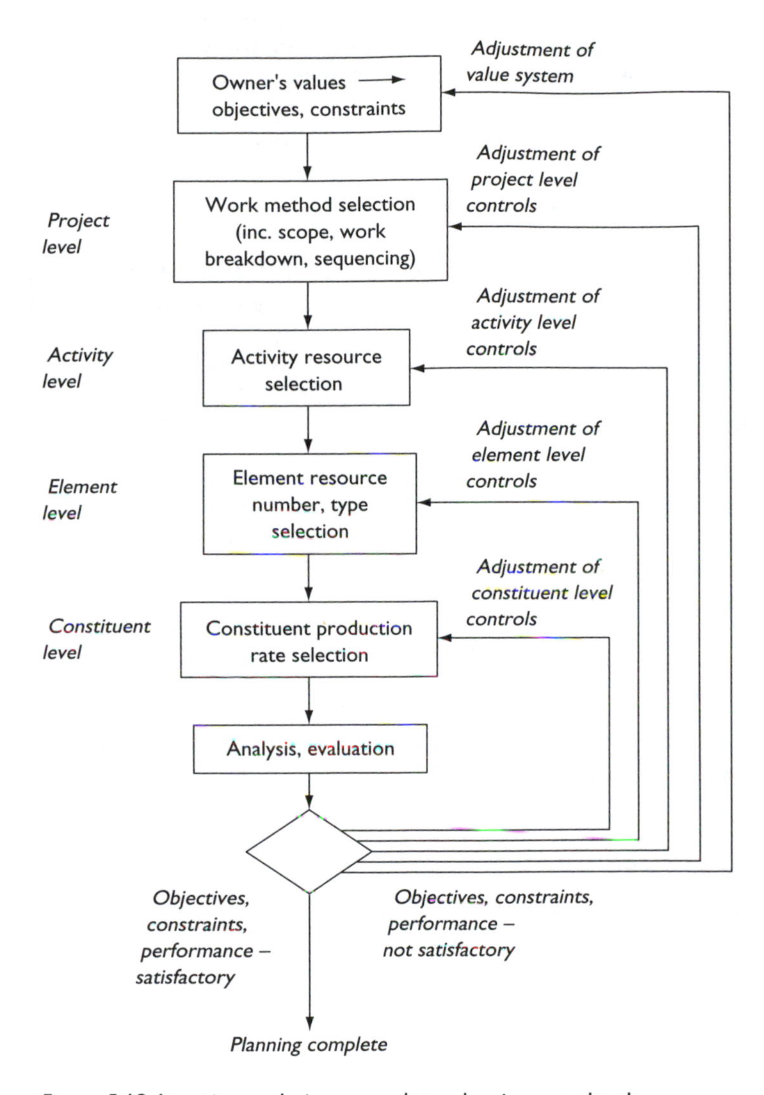

Figure 5.13 Iterative analysis approach to planning over levels.

Controls selected depend on the planner's experience and expertise, as well as the situation and some creativity. Some solutions may be better than others. Generally planners are after a satisfactory solution rather than some theoretically optimum solution.

If one approach doesn't work or doesn't work completely, then other approaches may be tried. Approaches are only limited by the planner's imagination. Knowledge of the project and the industry will help develop alternative approaches. Whether any of these approaches are possible will be determined by the project and the nature of the project work.

Any control adjustment, amongst other things, may:

- Lead to new cumulative resource (money) plots
- Alter the critical paths
- Lead to rescheduling of activities.

Value system adjustment may redefine the planning problem or, for example, lead to the removal of physical, technical or logic constraints affecting work.

Example

To assist an understanding of planning over levels, an example is presented. The example project chosen is one which most people should be able to visualise. The procedure in planning the project over levels is developed.

The simpler iterative analysis approach (compared to a synthesis approach) is presented.

End-product and project

An investigation report is required to be produced by a consultant. The project is the means by which this report comes about.

Objectives and constraints

The project objectives and constraints are from the viewpoint of the consultant undertaking the project (undertaking the work for an owner), but these could be expected to reflect the owner's values as well.

Suggested objectives might be:

- Minimum cost
- Maximum efficiency in work practices, including the usage of personnel.

Suggested constraints might be:

- Availability of expertise and people
- Duration upper limit
- Cash flow, perhaps requiring payment from the owner as soon as possible
- Fitting in with other projects.

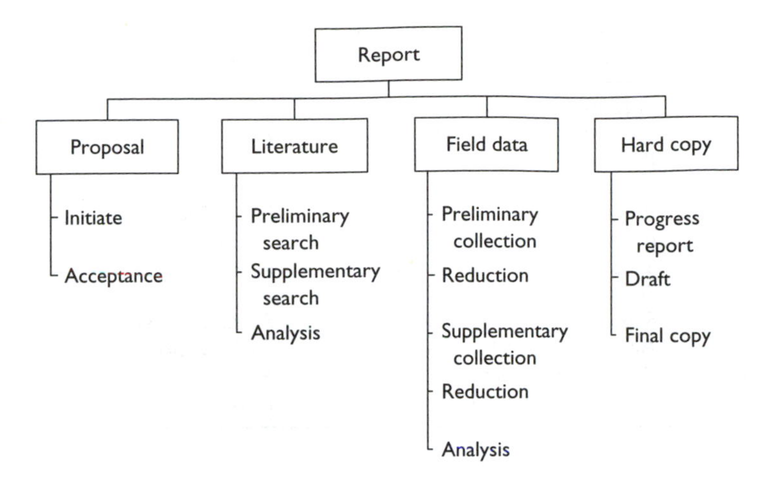

Figure 5.14 Possible work breakdown for the investigation report example. (Activity names have been abbreviated because of space limitations – work items are implied.)

Scope

Broad scope

The broad scope (of the project) is one of producing an investigation report. It involves developing a proposal, literature searches, the collection of field data and the production of hard copy.

The project proposes to use a number of people and their associated skills.

Scope delineation – Work breakdown

A possible breakdown to activities, not in any particular order, is given in Figure 5.14.

Project level

Activity interaction/sequencing

Precedence relationships establish the logic of how the project is to be put together. The following is one suggested way in which the work might be carried out.

Code	*Activity*	*Precedence*
Prop	Initiation of proposal	–
Accpt	Acceptance of proposal	Prop
Prelit	Preliminary literature search	Accpt
Predat	Preliminary data collection	Accpt
Prorep	Progress report	Prelit; Predat
Prered	Preliminary data reduction	Predat
Suplit	Supplementary literature search	Prorep
Supdat	Supplementary data collection	Prorep
Supred	Supplementary data reduction	Supdat; Prered
Anal	Analysis of literature and data	Suplit; Supred
Drrep	Prepare draft report	Anal
Firep	Prepare final copy of report	Drrep

Network formation

The network for this project can now be drawn. The network may be either an activity-on-link diagram or an activity-on-node diagram. The network looks like Figure 5.15(a or b). Figure 5.15a is an activity-on-node diagram. Figure 5.15b is an equivalent activity-on-link diagram.

Note that the network diagram is no more than the logic of how the work is to be carried out.

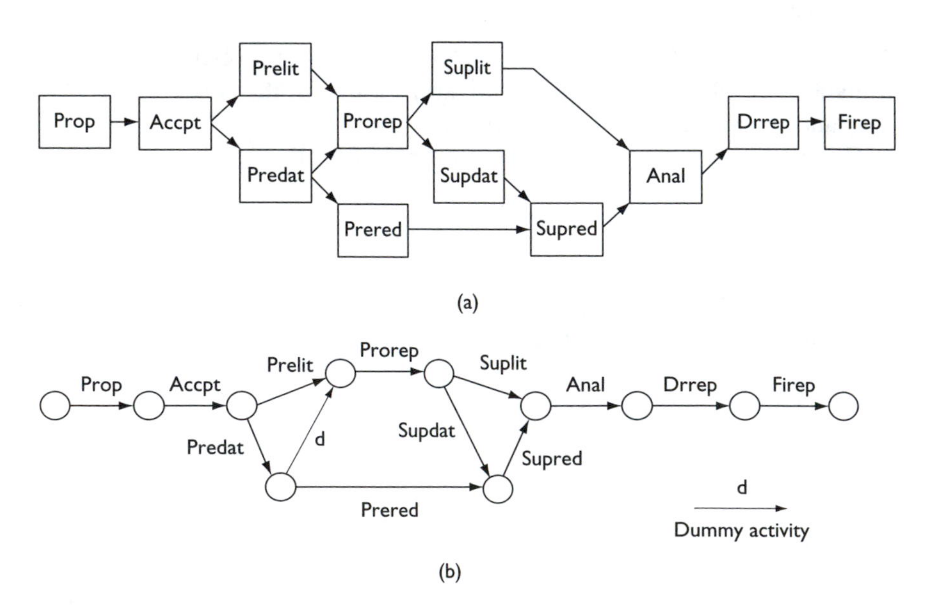

Figure 5.15 Example project networks. (a) Activity-on-node diagram; (b) Activity-on-link diagram.

Activity level

Activity duration estimates

Activity duration estimates imply that some assumptions have been made about resourcing and resource production rates. Duration, resourcing and resource production rates are related.

A possible set of duration estimates for the activities is as follows:

Code	*Activity*	*Duration (days)*
Prop	Initiation of proposal	10
Accpt	Acceptance of proposal	2
Prelit	Preliminary literature search	21
Suplit	Supplementary literature search	7
Anal	Analysis of literature and data	$2+4=6$
Predat	Preliminary data collection	25
Prered	Preliminary data reduction	6
Supdat	Supplementary data collection	10
Supred	Supplementary data reduction	3
Prorep	Progress report	5
Drrep	Prepare draft report	12
Firep	Prepare final copy of report	5

These values obviously depend on the skills of the estimator and the familiarity this person has with the work involved in the project.

Resource estimates

The resource requirements of 'professionals' is estimated to be as follows:

Code	*Activity*	*Professionals/day*
Prop	Initiation of proposal	2
Accpt	Acceptance of proposal	4
Prelit	Preliminary literature search	2
Suplit	Supplementary literature search	2
Anal	Analysis of literature and data	$1+1=2$
Predat	Preliminary data collection	3
Prered	Preliminary data reduction	2
Supdat	Supplementary data collection	3
Supred	Supplementary data reduction	2
Prorep	Progress report	1
Drrep	Prepare draft report	3
Firep	Prepare final copy of report	1

Money estimates

The following are estimates for costs to undertake the activities:

Code	*Activity*	*Cost/day ($/day)*
Prop	Initiation of proposal	1500
Accpt	Acceptance of proposal	3000
Prelit	Preliminary literature search	1200
Suplit	Supplementary literature search	1200
Anal	Analysis of literature and data	1200
Predat	Preliminary data collection	3000
Prered	Preliminary data reduction	1200
Supdat	Supplementary data collection	3000
Supred	Supplementary data reduction	1200
Prorep	Progress report	600
Drrep	Prepare draft report	1800
Firep	Prepare final copy of report	750

The agreement for payment (income) for the work is that $120 000 be paid after the completion of the progress report (Prorep), and a further $120 000 be paid on completion of the final report (Firep).

Time-dependent costs have not been included.

Element and constituent levels

The controls at the element and constituent levels are assumed fixed for this example.

Analysis

Network analysis

See Figure 5.16.

Bar chart

The information from a network analysis is readily transferable into bar chart and time-scaled network form (Figures 5.17 and 5.18).

Figure 5.17 shows the bar chart plotted for all activities starting at their earliest start times.

Time-scaled network

Figure 5.18 gives a time-scaled network.

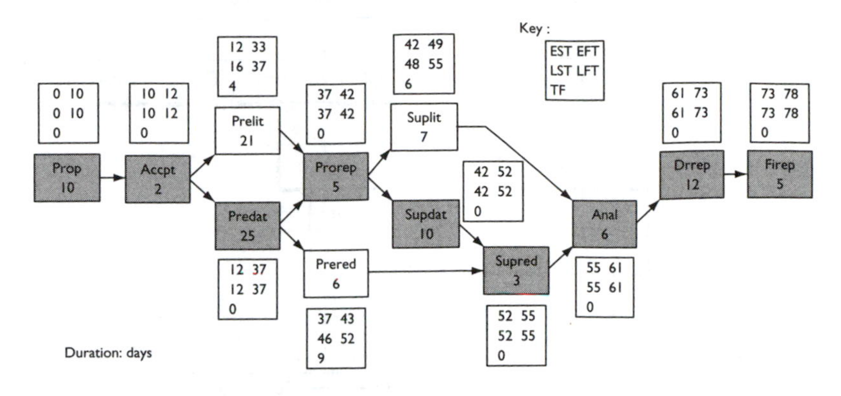

Figure 5.16 Example network calculations.

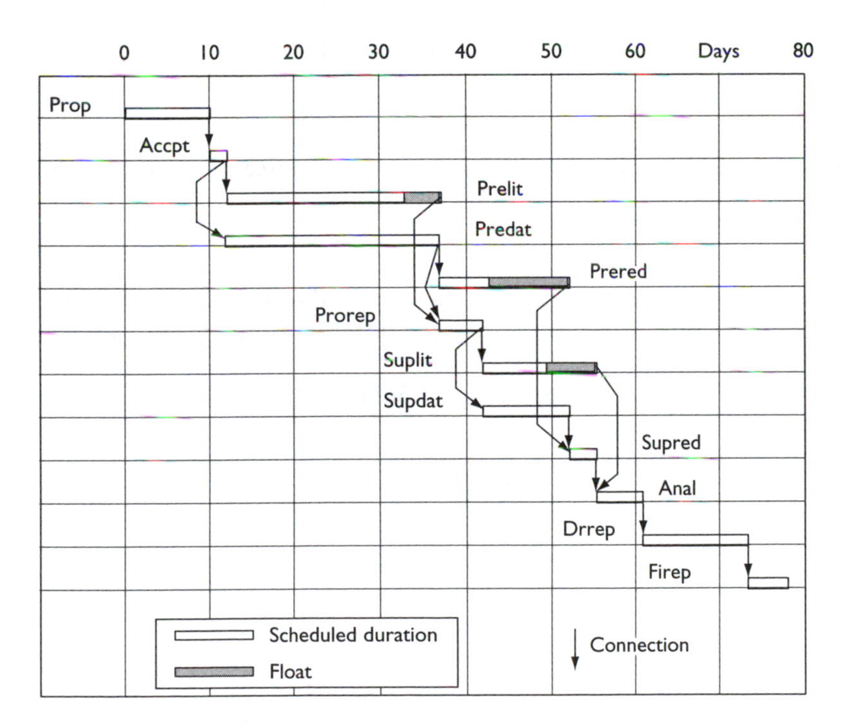

Figure 5.17 Connected bar chart for example project.

Calendar

For use by project personnel, the information is more desirably presented on a calendar allowing for stipulated work hours per day, weekends, public holidays, rostered days off (RDO), number of shifts per day and so on.

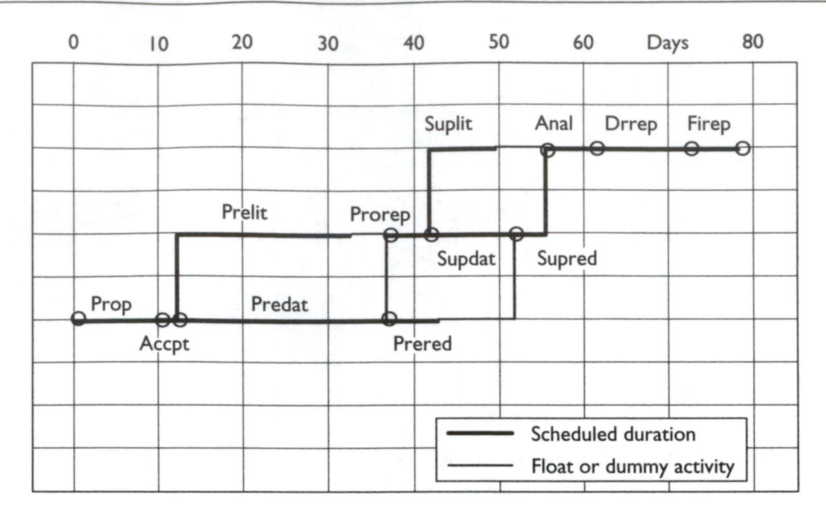

Figure 5.18 Example project time-scaled network.

Resources

Figure 5.19 shows the resource plot for the resource of 'professionals'.

Money

Plots of money requirements (costs) – expenditure plot, and a cumulative expenditure plot or S curve – over the duration of the project are given in Figure 5.20.

With income superimposed, Figure 5.21 results.

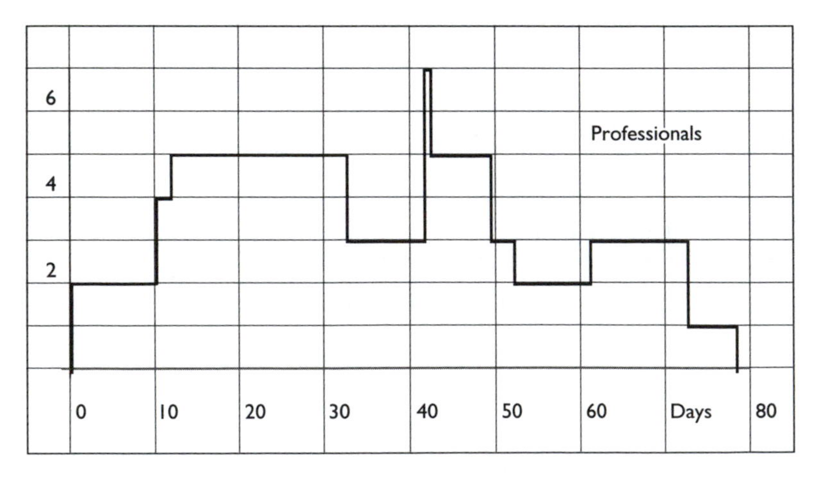

Figure 5.19 Example resource plot for project 'professionals'.

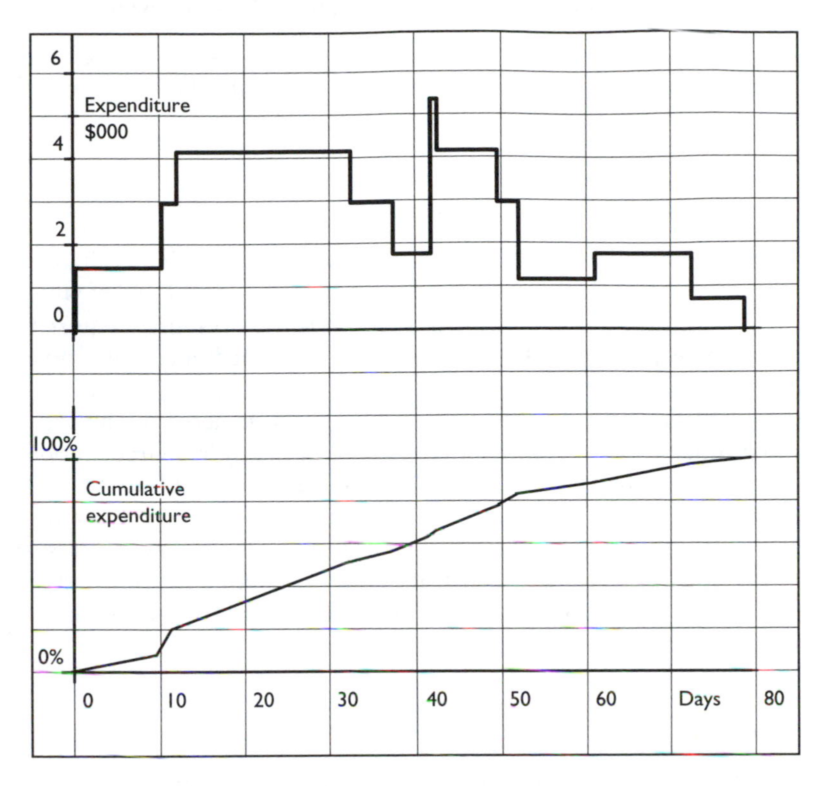

Figure 5.20 Example expenditure (upper diagram) and cumulative expenditure (lower diagram) plots.

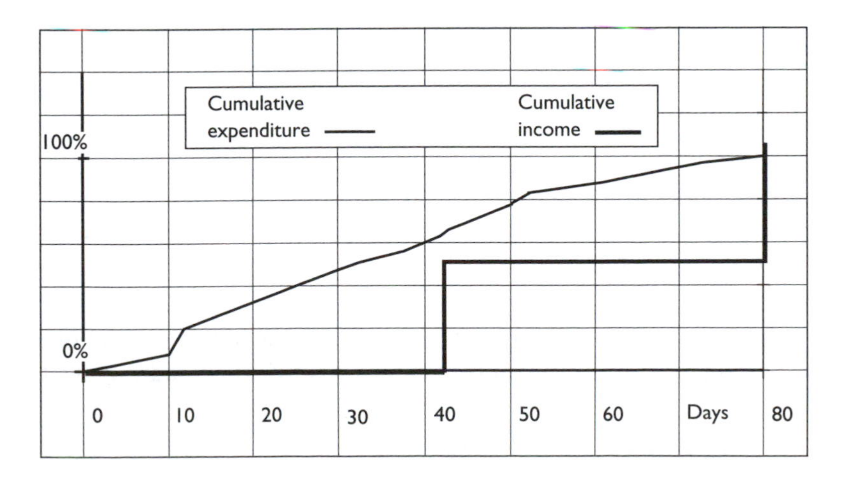

Figure 5.21 Example cumulative expenditure and cumulative income plots.

Evaluation against objectives, constraints and performance

The results of the analysis are now evaluated against the project objectives and constraints. How is the project in terms of schedule/production, resources and money?

Feedback – Adjustment of controls

The controls are adjusted at the relevant levels based on this evaluation.

For example, should the availability of professionals be fixed at five people, then something has to be done to reduce the peak resource requirement. Conventionally this is called resource constrained scheduling, but is no more than an adjustment of the controls.

As an another example, the resource usage as depicted in the resource plot may not be considered efficient. Conventionally resource smoothing addresses this problem, but again it is no more than an adjustment of the controls.

There may also be feedback to adjust the project owner's value system if the current objectives and constraints are unsuitable in some way. The objectives and constraints may also be modified if the analysis indicates something unusual.

Repeat

The whole process is repeated until the objectives and constraint satisfaction are to the liking of the planner. Experienced planners may only need one, or even zero, iterations. Inexperienced planners may need many iterations.

Exercises

1. Draw a tree diagram for the assembly of a machine or toy with which you are familiar. At the top of this diagram is the toy or machine. The next level represents subsystems; the next again level sub-subsystems and so on. The bottom level represents the individual nuts and bolts and discrete parts.

No two tree diagrams will be the same. It depends on the machine or toy and the way it is put together.

2. Assume that you are undertaking a project of holding a technical or public seminar. Some of the more important activities would involve those related to:

- Attendees (informing, handling queries, enrolling, payment, . . .)
- Venue (selection – size, etc., booking, refreshments and meals, payment, room preparation, audio-visual facilities, whiteboard, . . .)

- Speakers (selection, transport, accommodation, fees, . . .)
- Proceedings/seminar notes (writing, design, printing, binding, distribution, certificates, . . .)
- Promotion (market establishment, advertising, distribution, . . .).

All of this has to be done under various schedule/production, resource and cost constraints.

Develop a work breakdown structure for this project.

Consider now, the way a budget might be developed for this project. How could the budget format relate to the work breakdown structure?

How do you make duration estimates to carry out each of the earlier mentioned activities?

Consider now the personnel involved. These might include:

- Speakers
- Company director
- Project manager
- Assistants to the project manager
- Attendees.

Relate the organisational structure that you might use to the work breakdown structure that you have just developed.

3. How do you know what is the lowest level to decompose a project?

4. A work breakdown structure (WBS) is chosen to suit the needs of each project. It is an accepted way of representing how a project is broken into finer components. However, it is possible to not be systematic and describe the work in an unstructured list.

What is the difference between an unstructured activity list and a WBS? Why is the latter preferred?

5. Consider a project associated with a social or sporting group to which you belong. Develop a work breakdown structure for the project.

Draw a responsibility chart/matrix (Chapters 3 and 7) concerning all the project participants.

Draw an organisational chart identifying all the people involved.

Relate the work breakdown structure to the responsibility chart/matrix and organisational chart.

Relate duration, resource and cost estimates of the project to the work breakdown structure.

6. A work breakdown structure subdivides a project into hierarchical units of subprojects, . . . , work packages, . . . , activities. The end result of the work breakdown process is a collection of work units or activities of relatively short duration and of manageable size.

Each activity is of a size to facilitate estimating the number and type of resources, money and duration that are necessary to accomplish the activity.

Each activity is a meaningful job for which individual responsibility can be assigned.

Consider a project associated with a social or sporting group to which you belong. What is the lowest level at which estimates of resources, money and durations can be made?

Chapter 6

Replanning

Introduction

Replanning is carried out according to project stages or at discrete points in time. Replanning is no different in form to planning. Replanning updates the controls based on actual performance, the current project state and some forecast of future influences, as the project progresses. Information assisting the replanning problem at any stage is better less far into the future. The (re)planning problem (for the remainder of the project) becomes better defined as the project progresses. As well, the planner's experience and understanding of the project improves. The level problems repeat for each stage, but with updated information.

During project execution, attention centres on keeping up to date with the effects of changes, delays and unforeseen events which on many projects appear inevitable. This implies continual replanning. A good plan allows for some flexibility and can be adapted to changing circumstances that may occur during a project. However, attempts are made to foresee potential influences in upcoming stages, and controls (immediate or contingency) developed for them much like in risk management (Carmichael, 2004).

Planning starts with broad assumptions which are refined as the (re)planning progresses. Planning always proceeds from the general to the detailed. To attempt to plan in great detail right from the beginning ignores the reality that future events are difficult to predict with certainty and outcomes never eventuate as intended. As the project progresses, more information becomes available, 'unknowns' become 'knowns', and the (re)planning can become more detailed and more accurate.

The ability to influence the outcome of a project could be expected to diminish as the project progresses (something like Figure 6.1). Planning carried out early in a project accordingly could be expected to provide the best opportunity to determine the direction, cost and so on for the project. As the project takes shape, the opportunities disappear. Early, decisions could be expected to have the greatest effect. Conversely, changes made later in a project are viewed as more costly; it is considered important to get it right at the start. In Figure 6.1, trends only are shown and the location of

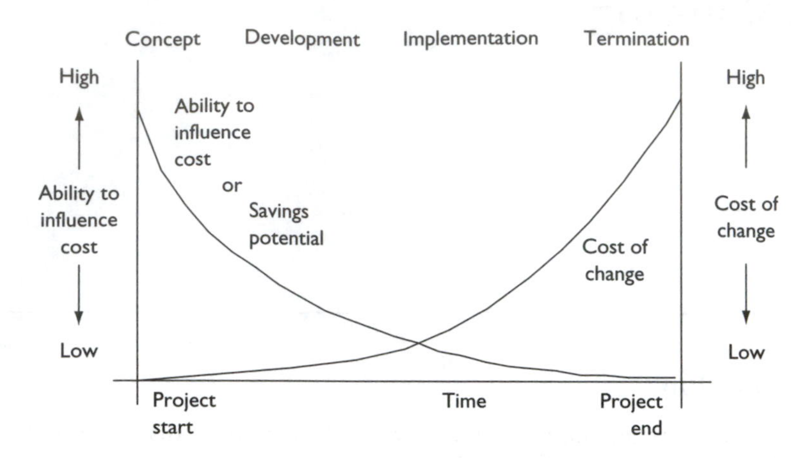

Figure 6.1 Anticipated cost influence and changes over a project's duration (Carmichael, 2004).

the intersection point of the trend lines is not important. How data could be obtained to substantiate the widely held beliefs of Figure 6.1 is unclear (Carmichael, 2004).

Planning, monitoring, reporting and replanning

Planning establishes a project 'baseline' or 'target' performance. Disturbance takes the project behaviour away from the current baseline.

As one precursor to replanning, the current output (performance, state) of a project is monitored and the information obtained converted into reports on which management decisions (selection of controls) can be made (Figure 6.2).

Monitoring involves measuring or observing the output or performance of a project.

Reporting is in absolute terms and terms relative to planned baseline or target performance (variances – the differences between planned and actual performance). Like programs, reporting is made appropriate to the level of management concerned. Information desirably is presented in a simplified and easily understood form. Reporting specifics can take various forms.

Replanning incorporates updated estimates of future influences, and uses the current project state as initial conditions. Variances are used to guide the choice of which controls need to be adjusted to correct the path of the project. Replanning gives an updated project baseline or target performance. Disturbance takes the project behaviour away from the current baseline. Monitoring and reporting follow, and the cycle continues throughout the project.

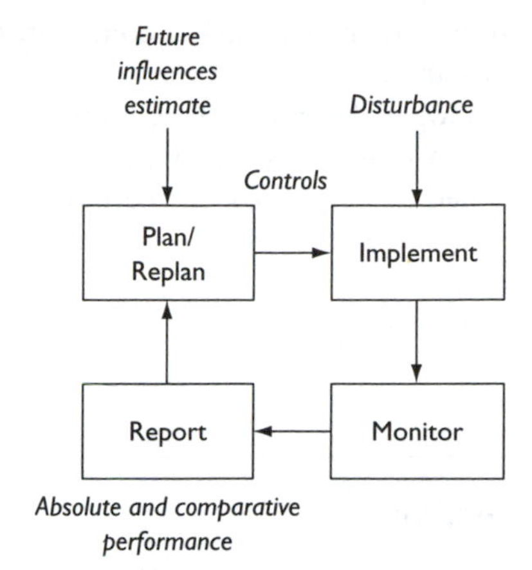

Figure 6.2 Plan–implement–monitor–report cycle.

Generally the greater the frequency of monitoring, it could be expected that the project will more likely perform close to that desired. Continual monitoring and feedback of project progress is necessary in a changing project and environment (reflected by disturbance), else replanning may reduce to something akin to crisis management. (Project) scope also rarely stays constant.

Planned-for project states may not be reached because of delays, weather disruptions, industrial disruptions, variations (design changes, owner priority changes, . . .), resource variability (within and between activities) and so on, even though allowances for these matters may have been included in the original planning. All these things are collectively referred to as disturbance.

In principle, replanning is no different to planning. It is planning carried out subsequent to any original planning. As such, all the discussion on planning applies to replanning.

Systematic, disciplined replanning is recommended practice.

Performance measures

Performance measures are gauges by which the success or otherwise of a project is measured. A performance measure is in the lay usage sense of the word 'objective' (mentioned in Carmichael, 2004).

Generally a project has a planned-for performance (resource usage and production over time; this may be translated to cost), and actual performance is compared with planned performance. One measure of a project's success may be in terms of the difference between this planned and actual performance. (But comparisons between actual performance and other performances, for example industry standards/benchmarks, competitors' performance or previous projects, are possible.)

Performance measures tend to be developed later in projects (after the initial planning has been done); objectives are developed very early.

A performance measure might be termed a 'key performance indicator' (KPI) or similar, but all of these types of terms are used very loosely by most people. The terms are used frequently by project personnel because they sound good and impress, but lack precision.

Case example – Powerline construction

This case example looks at compression practices to meet a powerline construction project's deadline, when factors external to the project, such as long delays in obtaining government approvals, and unexpected wet weather arose.

This project started with a route selection study, environmental impact statement (EIS), design, materials procurement, survey and the letting of the construction contract. Governmental approval was not obtained until a year later, which meant that there were two months less available than originally estimated to construct the powerline, weather permitting.

Crucial to the success of the project would be close and constant monitoring of the construction activities to ensure that the project deadline was met.

The tender documents for the line construction were distributed just prior to obtaining the government approval to proceed with the project. So the owner was able to interview the final contenders and advise them of the project's requirements, with emphasis being placed on the project's deadline and their proposed work schedules to complete the work as required. The actual construction was split into two parts, a section of the line on a mining lease area, and a section off the mining lease area.

The contractor was given the responsibility for arranging and coordinating the delivery of the concrete poles with the pole supplier. One of the major uncertainties of the whole project was the delivery of the concrete poles to site, in this case to the pre-drilled hole sites, because of the terrain.

At the same time, other infrastructure work for the mine was being conducted in the vicinity of the line, namely the laying of a pipeline in a main mine access road reserve. The road was also the main access route for the powerline construction, and this made delivery of the poles very awkward in the majority of instances. This was an item which was not known about or considered in the original project plan. This route was also the main transportation route for the delivery of equipment to and from the mine site, and this ensured this particular road was always busy.

The successful contractor established itself on site very promptly. The first month or so went very smoothly, except for obtaining some environmental approvals for benching work on the mine site, but this was quickly cleared up at one of the regular weekly meetings with the owner.

Then Murphy's Law came into play; it started to rain. As a result of this, the poles could not be delivered to site. This meant that poles were being dumped at specified sites, and additional cranage and transportation was required. The project started to fall behind schedule, and all stakeholders were starting to become concerned.

In conjunction with the contractor, the owner started to look at what could be done to obtain some slack in the program to ensure the project would meet the required deadline.

The following items were identified as areas where the project could be progressed or compressed:

- The contractor's original proposed work schedule was looked at, and it was discovered that it was one work team short. This was due to the fact that it had not started pipe stringing as yet. This was rectified and the additional team was put in place.
- Having looked at the proposed schedule, it was mutually decided to vary the schedule and start work in those areas that were accessible even though some other planned areas were inaccessible due to the wet weather. This meant some staffing re-arrangements, but it seemed to work effectively.
- Lengthened work hours were suggested. However, this was thought to be a major problem, because part of the government approval conditions stipulated the working hours and days – Sundays were not included and the work hours were to be between 7.00 am and 5.00 pm.

  Approval for extended working hours was obtained. This meant the contractor could utilise the additional hours, and the contractor was also allowed to operate on Sundays provided this was not done in the vicinity of the local town.

The concern was for the impact of all these items on the estimated project cost. The other major concern was that the contractor had about two weeks of delays due to wet weather, which could have meant that the deadline was not going to be met anyway.

The canvassing of options was handled openly among all parties concerned. This enabled timely and efficient solutions to be identified and allowed processes to be put in place to resolve these issues.

The additional cost to the owner was negligible compared to the cost of not meeting the deadline, and so the contractor was given the go ahead to work the extended hours.

The contractor still struggled with the resources that it had and inevitably towards the end, additional resources were required to ensure that the project deadline was met. In the end, the project just met the deadline, due in part to good luck rather than good management. The owner was satisfied and the additional costs were negligible compared to the overall cost of the project.

Case example – Documentation for lng plant expansion

This case example involves the selection of controls in a project involving the development of documentation related to the expansion of a liquefied natural gas (lng) plant. The same contractor who had successfully designed and built the original plant undertook this documentation project. The owner had a prominent involvement in the project, electing to have an owner's team reside in the same building as the contractor's team. This was aimed at improving communication between the contractor and the owner, as well as to increase the efficiency of the approval process. The project was initially on a tight schedule and ran in parallel with finding purchasers of the product from the expanded plant.

There were many sub-teams involved in the project, each specialising in an area of expertise such as electrical and instrumentation, fabricated equipment, piping, special equipment, rotating equipment and process engineering. A number of technical groups were combined to form a larger group under the direction of one project engineer. The fabricated equipment group, special equipment group and rotating equipment group were all under the direction of the same project engineer. It was the project engineer's responsibility to record the progress made by his particular technical groups each week and report this to the planners.

At the beginning of the project, one of the first tasks that the lead engineers of each technical group was required to perform was to give estimates of hours to finish each document. (The [project] scope was to produce a set of documents that would detail the requirements of the expanded plant.) This information was charted as an S curve, which formed the basis of all monitoring, reporting and replanning activities. The hours were approved by the owner and allocated to the technical groups. Each week the project engineers would report the progress of their groups to the planners and the hours expended would be plotted on an S curve so that the difference between planned and actual hours could be seen. At one stage, the rotating equipment group had fallen behind schedule (but had not expended the budget hours to get to the stage they had reached), while the special equipment group was ahead of schedule and hours expended. In order for the rotating equipment group to get back on track, they needed more resources. The owner, however, had a limit on the resources to be used on the project, and so to eliminate the need for additional resources, the project engineer rearranged resources for a short period. One engineer from the special equipment group moved to the rotating equipment group for one week and performed tasks that relieved that group's workload. The rotating equipment group was able to move back onto schedule without employing additional personnel. It was important, however, to ensure that all hours worked by the special equipment engineer on the rotating equipment tasks were allocated to the correct cost code. For costs to have gone to the special equipment budget would have been a misrepresentation of actual costs spent in each area, and hence further planning would have been incorrect due to false information.

The planners would track progress each week through planning meetings with the project engineers, where verbal reports on progress were given. A group being behind schedule, however, was not the only problem to them. A group being ahead of schedule was just as troublesome to the planners. The reasons for this were twofold:

- The schedule was based on a flow of information. The special equipment group relied on information from process engineers. The process group had a schedule to deliver technical information and if the special equipment group was ahead of schedule, there would be a wait involved for technical information from the process group.
- The owner would review the schedule each week. If many groups were seen to be ahead of schedule and others behind schedule, it would show poor planning and a poor allocation of resources.

> This would reflect adversely on the contractor as a professional engineering consulting firm.

The planners would give the project engineer instructions on where each group needed to progress in the coming weeks, and then the project engineer was responsible for reallocating personnel to ensure that all groups stayed on schedule.

A standard template, for every document, determined the quality of the documents. This, however, did not exist at the start of the project but was established with input from the groups that were first to reach the stage of requiring a template. This showed signs of poor planning because many groups were held up at the beginning by first having to create the standard template for documents, and then waiting for the owner to approve the layout. This is a task that should have been completed during the initial stages of the project. The project engineer was responsible for requesting a further budget allocation of hours for those groups who were involved in the development of the standard format, which was not initially identified as a task for the groups.

There were various forms of control that the project engineer utilised on the project, the main form being the reallocation of personnel to suit the scheduled requirements. The other form of control was to request from the owner, additional hours for work performed, hours that were not included in the initial planning.

Monitoring

General

Monitoring forms an essential part of the (re)planning–implementing–monitoring–reporting cycle. It encompasses the tasks of collecting and recording information concerning the behaviour or performance of the project; this will reflect how the project is going, at any given time, in terms of production (work completed, attainment of milestones, . . .), resources (money) used and so on. This information is used in comparison with the planned performance to influence the selection of updated controls.

Changes in scope, that is changes in the extent or nature of the work itself, are additional considerations commonly monitored. Changes in scope also feed into the replanning.

The Pareto rule could be expected to apply. It may only be necessary to monitor 20% of project performance at any one time in order to get an indicative feel for how the project is going.

Data on performance is collected at the lower project levels and summed to give performance at higher project levels.

Frequency of monitoring

The emphasis in monitoring is on providing information on a timely basis. The frequency of monitoring is determined by considerations such as:

- Duration of the project
- Value of the project
- Current project work volume; how many activities are underway at any time
- How fast the project is progressing
- Degree of smoothness or number of troubles or complications on the project; degree of difficulty of implementation
- The project stage or phase
- Resource availability to undertake the monitoring
- The importance of the current activities
- The stakeholders' expectations
- Organisation or industry norms or culture
- The level in the organisation using the information.

Data collection

Monitoring involves data collection. Data collection practices may vary between industries, between organisations, between individuals, between in-house projects and outsourced projects, between owner and contractor and between delivery methods. Usually, however, information on the following is required.

For (re)planning purposes

- Production
  - Equipment, people (labour) and materials
- Resource usage
  - Arrivals, departures, duration, costs, number (quantity) and type
  - Equipment and people (labour).

Both are related back to activities and events for progress determination purposes.

Changes in scope are also monitored.

For project management purposes

- Contracts – performance
- Project overheads
- Quality
- People behaviour.

Computers assist in the storage of the huge amounts of data that are collected as well as in the reporting function (creation of legible reports of performance in absolute terms, and reports comparing planned and actual and showing slippage, overruns and underruns). Forms for the manual collection of data are designed with the view to transferring the data to a computer. Data consolidation takes place before reporting. Information on resources is converted to a money unit for cost reporting and replanning purposes.

Example – People (labour)

Data on labour is collected using standard forms such as Figure 6.3. In association with the labour data, data on production is also recorded such as by the use of forms illustrated in Figure 6.4. The person in charge of the workers is best placed to complete such forms.

Information from such forms can be used to establish production and can be used to assist in estimating remaining durations to complete activities currently in progress. The latter use facilitates replanning.

Data is regularly collated and developed into status reports. Labour costs are related to the production.

Project **Date**								
Occupation	**Employee**		**Activity Code**	**Regular**		**Overtime**		**Amount**
	Name	**No.**		**Hours**	**Pay rate**	**Hours**	**Pay rate**	

Figure 6.3 Form for collecting data on labour – daily 'time' sheet.

Project Date Activity				
Work item	**Measurement unit**	**Planned**	**Actual**	**Latest revised**

Figure 6.4 Form for collecting activity information.

Case example – Road project

In a road project, the first stage involved a bulk excavation activity of cut to fill. The operation used hourly hire plant from several companies. The spread of equipment involved was:

- *Group 1*: Scrapers, dozer, compactors, pad foot and flat drum rollers and graders.
- *Group 2*: Articulated dump trucks, excavator, graders, dozer and compaction equipment.

Costs on the project compared with the budget were usually reviewed on a monthly basis. The inherent variability of equipment posed a problem with uncertainty in productivity. Activities, using methods which could be susceptible to variable production in different ground and stockpile conditions, required tighter monitoring of costs. Costs were evaluated on a daily basis. If the production for the day was different to the planned production, methods could be altered. It was important to study daily cost data and format the data into production costs, to increase confidence in planning and forecasting of activities, for example a change over from articulated trucks and excavator to scrapers and dozer.

On any one day the two groups of equipment could be working on several different types of activities, so an overall measurement of productivity would prove to be inaccurate. As a result, a practice of docket submission by all operators of cartage plant (scrapers and trucks), including load counts, and also recording labour used, material moved and supervision, was implemented in order that the total overall cost could be assessed.

This method gave an estimate of the amount of material moved, which was reconciled against the actual survey of the volume moved, to give a close on the results.

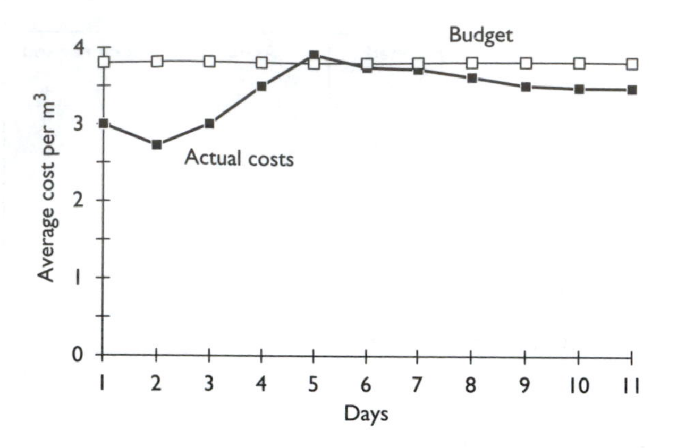

Figure 6.5 Example summary information.

An analysis of project production rates for the various cut to fill operations was undertaken, for both overall cut and fill, and for each spread of equipment. This last information could determine the efficiency of certain operations and detected whether the excavators and trucks were better suited to hauling from the stockpile or the cut, compared with scrapers.

Cost codes were used to break down the dockets and the whole of the information was entered on a spreadsheet to quickly analyse and determine project production rates.

The information was presented in a manner that could be easily read and understood by the project manager, engineers, superintendent and foreman. This enabled all those directly involved, to understand the performance of the work and enabled change or corrective action to be taken. Figure 6.5 illustrates the type of summary information.

Popular approaches to reporting

Introduction

Reporting project performance follows the monitoring. The frequency of reporting commonly matches the frequency of monitoring.

Reporting on performance at all project levels occurs. Each has its place. Much reporting practice has been built up over many years, and is naturally ignorant of any systems approach such as advanced in this book. However, all popular reporting can be related back to systems thinking.

Reporting may be to several levels of an organisation, possibly in correspondence with the work breakdown levels, and the information is tailored to the particular organisation level and its interests in the project. Thus, reports will differ in frequency and depth. Not all reports are used for planning; some are for information purposes only. Planners and managers close to the workface need detailed information about individual activities. Reports at this level are frequent. Senior management would require overview reports less frequently.

Reporting is carried out frequently enough to allow replanning to occur before the project can get too far 'off the track'. Timeliness of information is important. (See section 'Frequency of monitoring'.) The frequency of reporting also takes into account that there is an inevitable time lag between the recording of data and the processing of this data by computer into a form useable as a decision tool by the planner or project manager.

Reporting could also be expected to occur at project milestones.

Flow-on aspects to assist management

Reporting well done, besides providing a basis for the selection of replanning controls, gives rise to many flow-on aspects that assist the overall planning and project management. For example:

- It creates an awareness of the progress of concurrent activities, the relationship between activities, and of coordination troubles.
- It can provide early warning signals of potential troubles and delays.
- Unacceptable work progress and practices are detected early, allowing early correction.
- Important needs of the project can be distinguished.
- Owners and others not directly involved with the project work are kept informed of project progress; it provides knowledge of profit and loss throughout a project.
- Involvement from/by various stakeholders can be highlighted.
- Projects that are seen to be performing well can provide intrinsic motivation to the project team.
- Reporting can be considered part of the quality assurance process.
- Regular and thorough reporting can be regarded as a discipline.
- It provides a check on the accuracy of the planning.
- It has a use in resolving disputes that may arise.
- The estimating and planning of future projects benefit from the information.

Common reporting troubles

Common reporting troubles seen in practice include:

- No report
- Reporting for reporting's sake
- Too much reporting
- Too much detail, both in reports and in data collected; this creates an unnecessary burden on workers and supervisors who collect data; also it is easy to miss important information in reports that are too detailed
- Poor correspondence between the plan and that reported; the reporting ideally tracks information directly related to the project plan
- Self-deluding report, rose-coloured picture painted
- Falsified report; disguised status of project
- Assuming report
- Not directed at the end-user
- Lack of timeliness
- Presence of gobbledegook
- Poorly designed report.

A typical report

Comprehensive project reports contain information on:

- Resource usage, cumulative resource usage, resource usage variance. This is also converted to a common unit of money (cost).
- Production, cumulative production, production variance. This is also converted to information on schedule.

Also useful (and containing information on resource usage and production) are:

- Exceptions
- Milestones
- Updated project cost, cost variance and cost to go
- Updated project completion time, production/schedule variance and duration remaining
- Contingency/delay allowance usage and variance
- Incidents
- Reasons for possible ills or health, for example unexpected events that occurred. Suggested possible remedies for any ills.

This is in addition to graphical displays of progress:

- Variance plot
- A bar chart or time-scaled network

- S curves
- Cumulative production plots
- Cumulative delay plots.

And more general managerial/business issues:

- Any outcomes or spin-offs anticipated.

The reports may be for purposes other than replanning, for example, to take forward data to future projects, and to inform non-planners.

Where as-planned performance and actual performance are given, the difference assists the selection of replanning controls. Some of this information may be displayed in tabular form, but the graphical form tends to be more readily comprehended ('a picture is worth a thousand words').

Reports provide both the *positive aspects* of a project's performance and the *negative aspects*.

The term 'status report' may be used by some people.

A *variance report* indicates the difference between the as-planned and actual either in a raw unit (days, money, . . .) or as a percentage. This may be at the activity or project levels.

A *milestone report* indicates, for each project milestone, the planned value (days, money, . . .) for milestones yet to be reached, and completed values and variances for milestones reached.

An *incident report* lists any current or immediate-past problems, their potential impact on the project, and thoughts as to what corrective action may be used.

Exception reporting involves only highlighting those occurrences that are regarded as lying outside usual behaviour.

Cost reporting

Resource usage may be converted to a unit of money (cost), and cost reporting may involve all or some of, or be involved in:

- Variance reporting (budget minus actual)
- Milestone reporting
- Incident reporting
- Exception reporting
- Cost so far and cost to go.

A cost management summary something like Figure 6.6 might be used. In order to satisfy a project's cost considerations, management needs to be aware of the money spent on a project compared with the spending planned.

As invoices and progress billings are received, commitments are transferred to incurred costs.

Activity	Cost estimate	Committed	Remaining	Variations	Total (Forecast cost at completion)
Total					

Figure 6.6 Cost management summary.

S curves

S curves may be drawn for resource types (or money) as-planned and actual (Figure 6.7). A comparison of the two plots provides one means of reviewing the performance of a resource type (or money).

There is a need to relate the information in Figure 6.7 with information on production/schedule performance. For example, if the vertical axis was cumulative money/expenditure, Figure 6.7 shows spending less than planned. However, the diagram tells nothing of whether the project is ahead of or behind production/schedule. Ideally, Figure 6.7 should be related to a bar chart or time-scaled network showing progress to date. Alternatively, state space reporting (Figures 1.6, 1.7 and 1.8), or the idea of earned value may be used to relate expenditure and production/schedule.

Reporting resource (or money) performance on an S curve gives information only at the project level, not the activity or lower levels. Original data has to be examined to establish what is happening at the lower levels.

For the *money case*, the S curve shows cumulative money versus time, and may be called *cumulative project expenditure*. Where there is also income involved, for example a contractor being reimbursed for work done, the

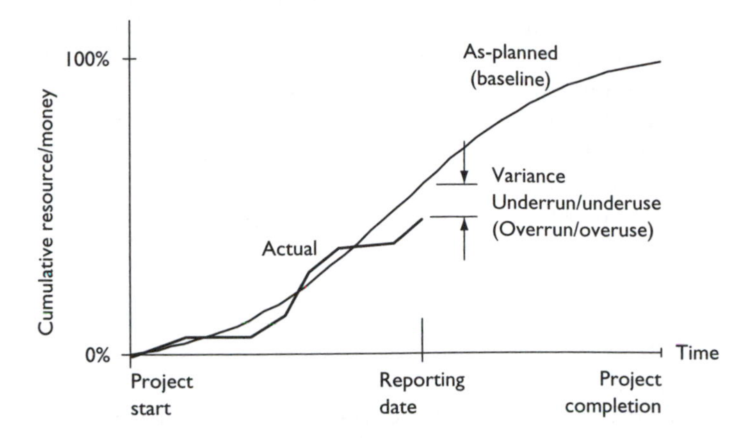

Figure 6.7 S curve or cumulative resource (or money) plot; as-planned and actual.

cumulative income plot can be superimposed on the cumulative expenditure plot to give something like Figure 3.17.

Milestone reporting

Milestones are shown on a network or bar chart in order that everyone is aware when particular important events occur, or when production should be accomplished by. Replanning may be necessary should the milestones occur late, or it becomes apparent that a milestone would not be reached by the desired date.

Milestones may be signed off as they are completed in a summary of milestones that can be consulted to determine the progress of the project. Milestone summaries may be used by people managing several projects in order to help minimise the degree of interference between projects.

Milestone reporting may be on resource (money) usage or on production/schedule.

One criticism of focusing on milestones as a measure of assessing whether a project is on track is that behaviour could be expected to follow Parkinson's Law where the milestone date has been set too conservatively. Placing greater emphasis on the final project completion date and lesser emphasis on milestone dates plays down the influence of Parkinson's Law. These comments relate to the case where penalties are attached to not reaching milestones as scheduled; where payment or incentives are attached to reaching milestones, behaviour will be different.

Those accountable for both estimating and achieving milestone dates quite naturally include some 'time' reserve in the estimate. Accountability only for achieving a project completion date could be expected to lead to a lesser total 'time' reserve in the project.

One suggested practice, if milestones are to be used as a focus for reporting (and no incentives are in place), is to include the idea of a buffer as used between activities in cumulative production plots and linear projects. Where the milestone separates project stages, a buffer is incorporated between the stages. This buffer reflects the conventional 'time' reserve in the milestone date estimate. Use of this buffer is then reported in the same way as delays. However, there is a view that tracking actual project progress against a minimum estimated duration and a buffer may increase reporting for little benefit.

Exception reporting

Exception reporting is analogous to practices used in statistical quality 'control'. There, readings lying outside upper and lower tolerance limits are regarded as matters for concern; readings lying within these limits are regarded as due to natural variability and not acted upon. See for example, the chart in Figure 6.8. In projects, only that information that is regarded as an outlier is acted upon.

Tolerance ideas might be transferred to S curves, such that an envelope of resource (or money) usage establishes the upper and lower limits (Figure 6.9). These limits may be, for example, ±10%. The project is allowed to proceed without adjustment as long as the actual performance lies within the limits.

Schedule reporting

What is termed 'schedule reporting' is reporting on production. The report of actual production/schedule performance is commonly presented as a tabular status report (for example, Figure 6.10 gives two possibilities). Reporting may be selective and only occur on critical or near critical activities.

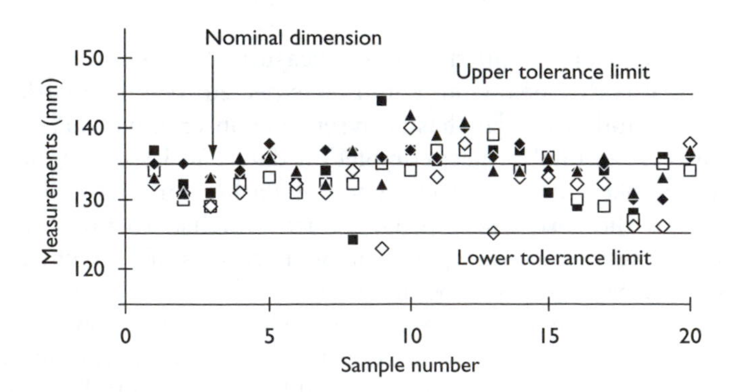

Figure 6.8 Example statistical quality 'control' chart.

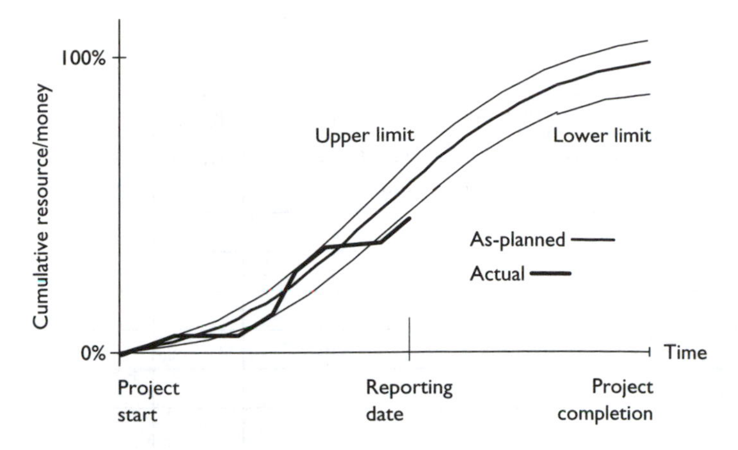

Figure 6.9 Limit envelope on an S curve.

In Figure 6.10b, the 'Comments' column records reasons for particular entries, anticipated future delays and so on. All activities – physical, administrative, delivery etc. – are included. This may mean more than one person completing the status report for any project.

The interpretation of the per cent complete column is important because some items, though technically or physically not complete, can still allow following activities to start. For example in building a house, roof tiles might still require grouting, yet at the stage where all tiling except the grouting has taken place this provides an adequate wet weather cover for the internal building trades work to start. Of course, this fact could have been taken into account in the scheduling by using an overlapping relationship between the activities of roof tiling and internal trades work.

The information in the status report can be transferred to the network. A time-scaled network or connected bar chart is more convenient for the recording of the progress of the project. Some colour coding or shading scheme is useful for distinguishing between the as-planned and actual performance.

Humour has it that: *The more you fall behind, the more time you have to catch up.*

Bar charts

Bar charts are commonly upgraded by drawing, parallel to an activity as-planned bar, a bar of length proportional to the amount of the activity production completed, commencing at the date when the activity actually started. Figure 6.11 shows two different drafting styles. A vertical line on

Project: Date:							
Activity	Estim. Dur'n	Scheduled			% Comp.	Rem. Dur'n	Comments
		Earliest Start	Latest Finish	Total Float			

(a)

STATUS REPORT PROJECT .... DAY						
Activity	Completed Activities		Activities in Progress			Comments
	Start date	Comp. date	Start date	Days in Progress	Estimate of days to Complete	

(b)

Figure 6.10 Example activity status sheet/progress report/form.

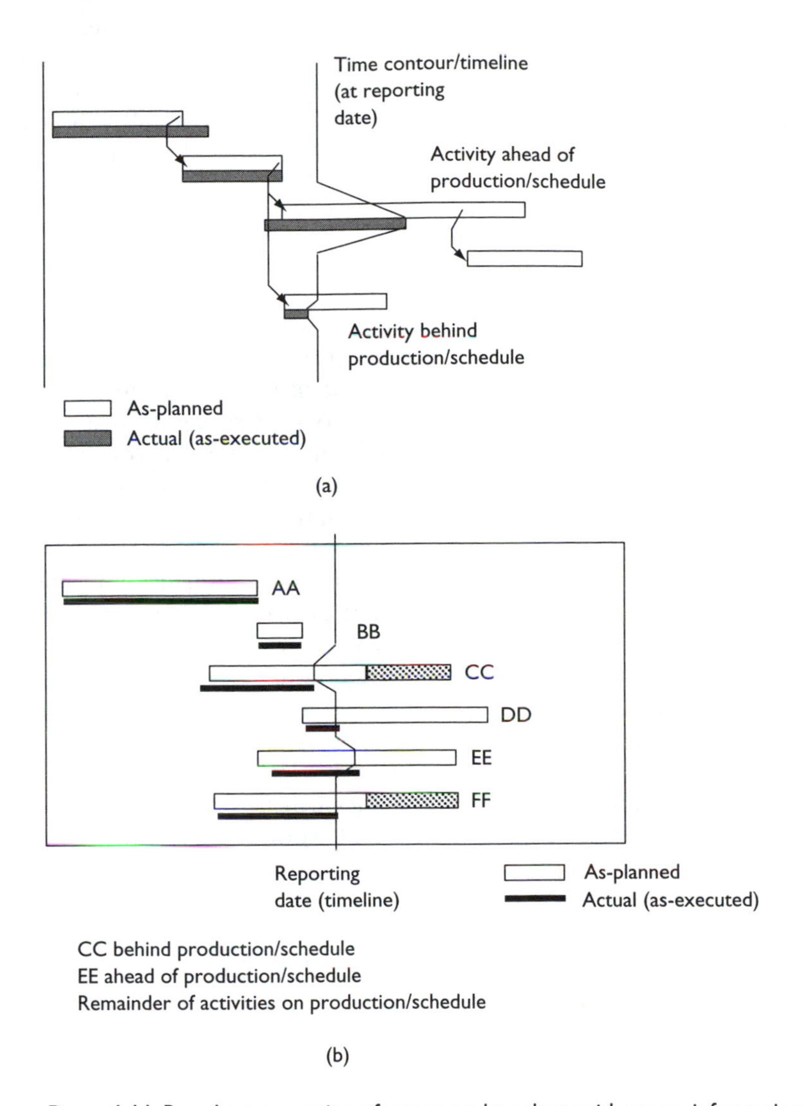

Figure 6.11 Bar chart reporting of progress; bar chart with status information.

a chart at any date indicates where the project and individual activities or production should be up to at that date.

Replanning (in an iterative analysis sense) will constitute at least a new network analysis leading to a new bar chart; the analysis is based on the status report, using the current status as the initial conditions for the analysis of the remainder of the project. However, if the intention is to stay with the original program then this new analysis is not necessary.

Time-scaled networks are treated similarly.

Reporting production/schedule performance on a bar chart gives information only at the activity level, not the project level. Hence, conclusions cannot be drawn from the bar chart alone as to how the project is performing with respect to production/schedule, though an overall 'feel' of how the project is going relative to planned production or original schedule can be obtained.

Factual networks

A factual network is also called an as-executed (as-built, as-constructed, as-implemented) program. It is a network (best done on a timescale) of as-executed activities indicating changes to the original program, including delays and variations, and agreed changes, cross-referenced to files on this information. It is a record of facts. The factual network becomes a permanent record of a project's progress. A factual network (as opposed to the original network) may not lend itself to analysis via conventional methods. The factual network also finds a use in attempting to resolve disputes involving delays, time period extensions, liquidated damages and related claims. It is similar in concept to an as-built drawing, in that it shows the actual durations to perform the various work items (Figure 6.12).

Work changes, including variations, alterations, extras, omissions and deductions, as well as delays can be argued on a more sound footing using a network as a basis. If a network is included as part of the contract, then all work changes and delays can be argued relative to this. Delays and work changes affecting non-critical activities can be argued differently to those affecting critical activities. Delays and work changes caused by the

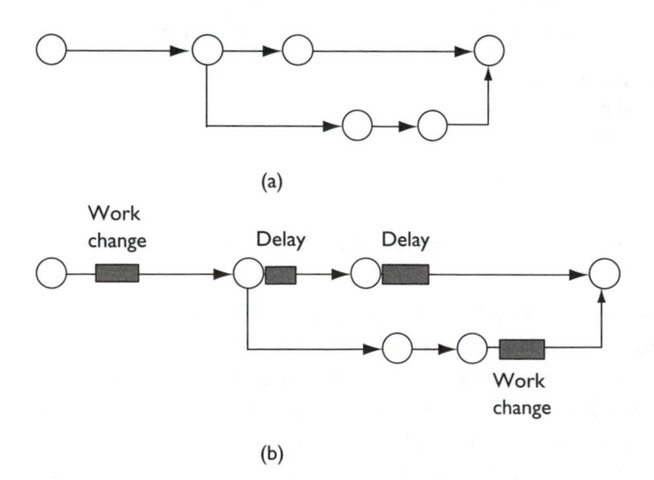

Figure 6.12 Example network showing changes and delays. (a) Network as-planned; (b) Network as-executed.

owner or the contractor, or by both the owner and the contractor, or by neither can be reconciled in principle far easier using a network of the project as a basis than without such a network. Should also direct and indirect cost structures form part of the contract, then the cost implications of work changes and delays can be established more equitably for both contract parties. Similarly time period extensions are placed on a more solid footing.

The analysis of the effects of work changes or delays is done by comparison with the network which was the original plan of the project. The factual network contains components that are the work activities as carried out, together with components due to changes and delays. This enables the responsibilities to be better ascertained for critical delays (delays affecting the overall project progress). Such calculations are often carried out manually although the computer coding of such an analysis is not a difficult task.

The factual network is supported with project records; the form adopted for recording such information can be flexible but would include space for cause, description, duration, other activities affected and comments.

Earned value

If a project is running ahead of production/schedule, then it would be expected that the actual cost would exceed the planned cost at that point in the project. Similarly, if a project is running behind production/schedule, it would be expected that the actual cost would be less than the planned cost at that point in the project.

Reporting project performance only on a conventional S curve (money) may give a misleading impression of the health of a project because it addresses only cost and not production/schedule. Earned value reporting attempts to address this; it is a reporting device at the project level where project performance with respect to both cost and duration is integrated.

The essence of earned value reporting is given here because of its use in industry. However, reporting in a *state space* is a *preferred* way of reporting (Figures 1.6, 1.7 and 1.8).

The earned value of production performed is found by multiplying the estimated per cent completion for the production by the planned cost for that production.

$$\text{Earned value} = \text{(Estimated) Per cent complete} \times \text{Planned cost}$$

This is the amount that should have been spent on the production at that point in the project.

Per cent complete and work packages

Estimating the per cent complete can be difficult to do with any precision, unless a large amount of effort is dedicated to the job. It can be approached approximately in a number of ways either at the activity level or by breaking the project up into small *work packages*, and then by following optional *rules*. Some example rules are:

- A 50–50 rule, where the activity/package is regarded as 50% complete if it has begun, and the remaining 50% is allocated when the activity/package is fully complete.
- A 0–100 rule, where the activity/package is only regarded as complete if it is finished. (This rule will understate the per cent complete because at any time there will be activities/work packages which are in progress.)
- Estimate per cent complete by 'eye' or judgement, depending on the nature of the work, and the usefulness to following dependent activities/packages of stating percentage completeness of anything less than 100%.
- Commercial computer packages may have their own peculiar rules.

Estimates made on work in progress will necessarily not be accurate. Work packages in progress will have part committed costs and part estimated/forecast costs. The total project cost will be a combination of estimates, commitments and actual costs.

Complete activities/work packages may be referred to as 'closed'; in-progress as 'open'; and unstarted as 'unopened'.

When an activity or work package is complete, its original budget may be referred to as being 'earned'.

The selection of the appropriate duration of a work package is unclear. The actual duration will relate to the overall project duration, the frequency of reporting and the individual planner's preferences. There is no definitive answer as to what duration to choose. Shorter duration work packages will provide more definiteness in estimating per cent complete because a greater proportion of packages will be either complete or not started; there will be a lower proportion in progress.

The plot shown in Figure 6.13 (an adaptation of S curve ideas) can be drawn and provides a basis for evaluating cost and production/schedule performance. The plot can alert management if variances are occurring that deserve further attention.

In Figure 6.13, three plots are shown – actual cost, planned cost and earned value. Reporting may give all three plots on one diagram as in Figure 6.13.

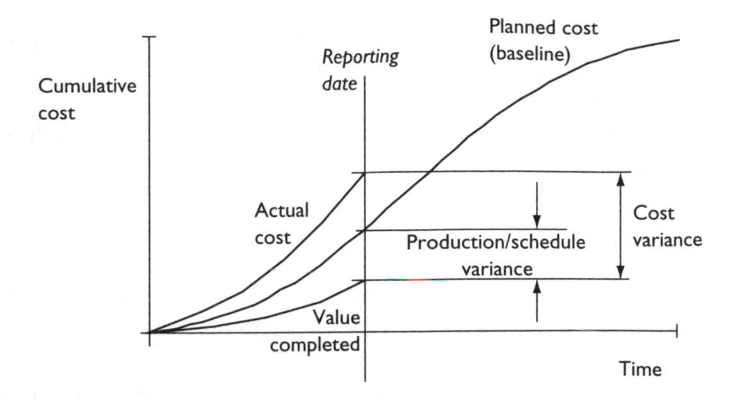

Figure 6.13 Example actual, planned and earned value plots.

Several types of variances can be identified in this figure. Two main variances are:

- Cost (spending) variance (underspent/overspent in a unit of money)

$$\text{Cost variance, CV} = \text{Value completed} - \text{Actual cost}$$

- Production/schedule variance (ahead/behind production/schedule in a unit of money)

$$\text{Production/schedule variance, SV} = \text{Value completed} - \text{Planned cost}$$

Note that the cost variance here is not the difference between planned and actual. The values of the variances are only as accurate as the estimates of per cent work complete.

Convention

The usual convention is for a variance to be *negative* if the project is *behind production/schedule* or *over budget* (Table 6.1), but other conventions will be found in the literature.

Table 6.1 Sign conventions on variances

	Negative (−)	*Positive* (+)
Cost variance	Over budget	Under budget
Production/schedule variance	Behind production/schedule	Ahead of production/schedule

Example

In Figure 6.13, at the reporting date, it has been estimated that the production planned to be done by that date is 50% complete. That is, value completed is 50% of planned cost.

This gives a negative production/schedule variance (indicating behind production/schedule), and a negative cost variance (indicating over budget).

Supplementary reporting

Information on cost variances and production/schedule variances can be put into more readily seen forms. Some examples are given here.

(1) Progress over the duration of a project might be plotted similar to Figure 1.7, but with axes of cost variance (CV) and schedule variance (SV).
(2) Progress over the duration of a project might be plotted similar to Figure 1.8, that is the change in cost variance and schedule variance with time.
(3) Dimensionless indices or ratios may be defined and reporting done in terms of these ratios or indices. The indices avoid the significance of variances being influenced by the size of the project. For example, a −$100, 000 variance on a $1 m project is more serious than on a $100 m project.

$$\text{Cost performance index (CPI) (CV ratio)} = \frac{\text{Value completed}}{\text{Actual cost}}$$

$$\text{Production/schedule performance index (SPI) (SV ratio)} = \frac{\text{Value completed}}{\text{Planned cost}}$$

The indices are particularly useful for establishing trends by plotting their values as the project progresses (Figure 6.14).

In the CPI formula, planned cost and actual cost are being compared. In the SPI formula, production performed and production scheduled are being compared.

Indices or ratios of 1 or 100% indicate that the project's progress is on target. Indices greater than 1 indicate that the project is doing better than expected. Indices less than 1 indicate unfavourable progress (Table 6.2).

The SPI indicates production/schedule efficiency. For example, an SPI of 0.75 or 75% indicates that only $75 worth of production has been accomplished for each $100 of production scheduled.

Such indices or ratios also permit comparisons between projects.

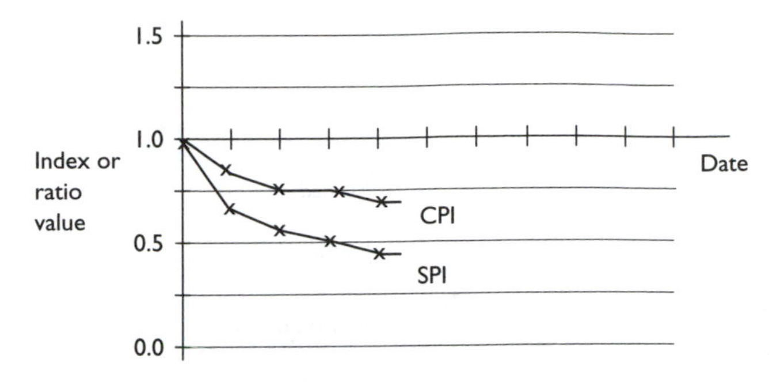

Figure 6.14 Example; plots of cost and production/schedule performance indices/ratios.

Table 6.2 Sign conventions on indices

	Index < 1	*Index = 1*	*Index > 1*
CPI	Over budget	On budget	Under budget
SPI	Behind production/schedule	On production/schedule	Ahead of production/schedule

Some earned value terminology not used in this book include BCWS (budgeted cost of work/production scheduled; also called PV – planned value), BCWP (budgeted cost of work/production performed; also called EV – earned value), ACWP (actual cost of work/production performed; also called AC – actual cost), STWP (scheduled time/duration for work/production performed), ATWP (actual time/duration for work/production performed) and cost/schedule 'control' system (C/SCS, CSCS, CS^2) because of their tendency to mislead. Earned value is sometimes called performance, and the alternative name, performance measurement (performance measuring technique, PMT), may be given by some people to the earned value approach; this terminology is also not used here. Some writers suggest, for reporting purposes, aligning work packages with project deliverables; however, since there is a lot of confusion by writers and practitioners over the use of the term 'deliverable', the term is best avoided.

Cumulative production plot

A cumulative production plot shows an activity's or project's production up to any time or distance; the slope of the plot represents the rate of production (Figure 6.15). Figure 6.15 can be either an activity level or project level representation. A number of variants on Figure 6.15 can be drawn (Figures 3.11, 3.12 and 3.13).

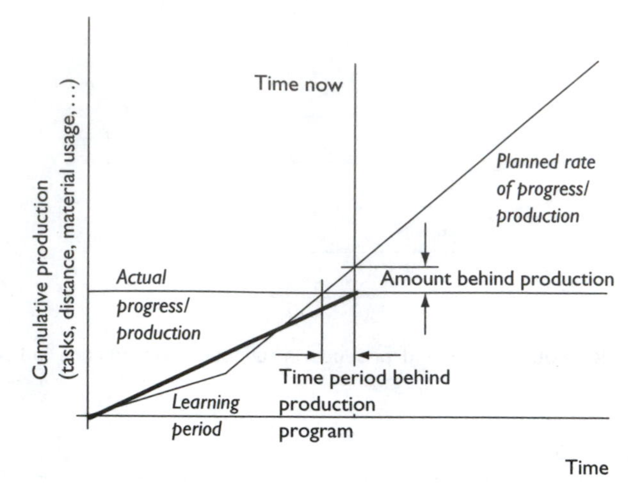

Figure 6.15 Example cumulative production plot.

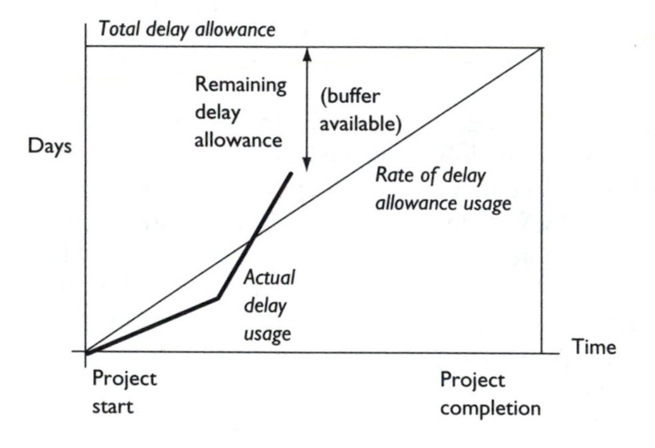

Figure 6.16 Example cumulative delay plot.

Delay reporting

Some projects include a delay contingency (production/'time' reserve) and monitor actual delays in production (Figure 6.16).

Case example – Land subdivision

This case example looks at one contractor's approach to monitoring and reporting on a land subdivision project.

At the estimating/planning stage of the project the construction work is subdivided according to a work structure that identifies/isolates similar work areas that are incorporated into the various scheduled items. This enables a comparison at a later date with actual costs. The selection of an appropriate degree of detail is considered important when assigning cost codes to work areas. A more detailed cost code breakdown would improve accuracy of the reporting but may tend to make the whole process unwieldy and not user-friendly.

Daily timesheets are kept that record all plant, labour and materials used for that day. These costs are proportioned to particular cost codes that are allocated to the different items within the work breakdown structure, for example earthworks, concrete work, and pipework bed, lay and joint. From the information contained on these sheets it is possible to accurately calculate the cost of any particular item and compare the actual cost rates to the tendered price rates. These sheets can then be totalled weekly and monthly to give the total costs for the month.

At the end of each month the costs that are established from invoices and dockets are totalled and printed out according to their allocated cost code. With the use of a spreadsheet, the cost for the month is compared to the budget for the month. From this, forecasts of budget overruns and underruns can be made and where possible, action taken to remedy any troubles.

Case example – Civil construction

Major items established at the commencement of the construction of a wharf-ocean retaining structure project to assist with replanning included:

- Construction program
- Cost centres based on construction program activities
- Budget allocation for cost centres
- Cash flow diagram for the project
- Budget summary for cost centres
- Personnel register
- Internal/external plant registers
- S curves
- Activity production rates.

Construction program

The construction program showed the sequence of activities and duration of each activity, and assisted in the estimation of manpower and plant required at any stage, and was used to indicate whether the project was behind or ahead of original schedule.

Cost centres
Major or summary activities, which separated the project into geographical areas or construction methods, were selected from the construction program. A cost centre was then allocated to each activity, and this gave a direct association between each activity and the budget allocated to that activity.

Budget allocation
Once a cost centre had been allocated to each major construction activity, resources were allocated to cost centres so as to estimate the cost to complete each activity.

Account codes
Account codes chosen were:

10 Labour
20 Subcontractor
30 Permanent Material
40 Temporary Material
50 Internal Plant
55 External Plant
60 Fuel/Repairs
90 Overheads

Resources were allocated to account codes. This was converted to costs per month and then multiplied by the activity durations to give cost-to-complete. The durations were taken from the construction program.

For recording of committed costs, purchase orders were filled out for every dollar spent and a cost code allocated to that purchase order. Details of purchase orders were entered into a costing program.

On receipt of invoice, delivery docket, purchase order and invoice were matched, authorised for payment and sent to head office.

Cash flow diagram
The cash flow diagram was prepared at an early stage in the project to assist the area manager in the management of cash flow for the office.

Budget summary
The budget summary gave a lot of important information on one sheet of paper because it indicated the project's financial status such as cost-to-date, cost-to-complete, budget and trends positive or negative.

Each month, the cost variance was recorded. Any trends in this variance were recorded and action taken, if necessary.

Personnel register
The personnel register provided the exact number of employees allocated to each cost centre, and therefore was very useful when allocating labour costs.

Internal/external plant registers
The plant registers were similar to the personnel register in that they gave information of exactly what plant was currently allocated to each cost centre. At the beginning of each new reporting period, a purchase order was written for every plant item and a cost code allocated, giving a committed cost for the month.

S curves
For each cost centre of a construction activity, an S curve was established to closely monitor progress. The plots had to be updated to account for increases in quantities or approved time period extensions.

Activity production rates
Activity production rate information was used for pricing and the programming of the project. Activity production rates were also used for the pricing of variations during the construction period.

Replanning representation

Introduction

Planning uses open loop control incorporating estimates of future influences. This gives a project 'baseline' or 'target' performance. Disturbance takes the project behaviour away from the current baseline.

As a precursor to replanning, monitoring and reporting of project performance gives project status in terms of absolute performance and performance relative to the baseline (errors or variances).

Replanning uses open loop control incorporating updated estimates of future influences, and initial conditions equal to the current project state. Error control ideas are used to guide the choice of which controls need to be adjusted, but are not used in an automatic closed loop control sense in current replanning practice. Replanning gives an updated project baseline or target performance. Disturbance occurs. Monitoring and reporting follow, and the cycle continues throughout the project.

Replanning is, in principle, no different to planning. Replanning controls are no different to planning controls, and involve the selection of method, resources and resource production rates. However, note that changing the work method may involve the development of a new network and associated flow on effects.

Summary

Project performance (the state $x(k)$) is monitored on an ongoing (usually regular) basis, compared with planned performance and, if necessary, corrective action is taken to send the project in a desired direction (on a desired state trajectory). The controls bringing about corrections may be in the form of method (including sequence), resource (or money) and/or resource production rate inputs to the project. The controls are chosen to extremise the objectives while satisfying any constraints present. *Applying changed or different controls amounts to replanning.* Replanning comes up with revised values of control variables, $u(k)$, for the remainder of the project – $u(k)$ has multiple parts (that is, it is a vector quantity of control variables) containing information on method, resources and resource production rates.

Generally it can be said that prior work done constrains future (updated) planning controls. Hence the project constraints are changing as a project progresses. The flexibility in choice of controls decreases with time.

The following representation is given for the case where the project duration remains constant. A similar representation could be given where the project duration is allowed to change.

Notation

The notation used in the following is:

N	number of periods/stages
k	period/stage counter, $k = 0, 1, 2, \ldots, N-1$
$x(k)$	planned-for state after period/stage $k-1$
$x(0)$	initial state
$x(N)$	final/terminal state
$xA(k)$	actual state after period/stage $k-1$
$x'(k)$	revised planned-for state after period/stage $k-1$
$var(k)$	variance; difference between planned-for state and actual state after period/stage $k-1$
$u(k)$	control at period/stage k
$\zeta(k)$	disturbance; anything that prevents $x(k+1)$ being obtained exactly.

Baseline

The (planned) baseline is obtained by solving the deterministic optimal control problem from project start to project end (or the equivalent iterative analysis version). Future uncertain events/influences are converted into

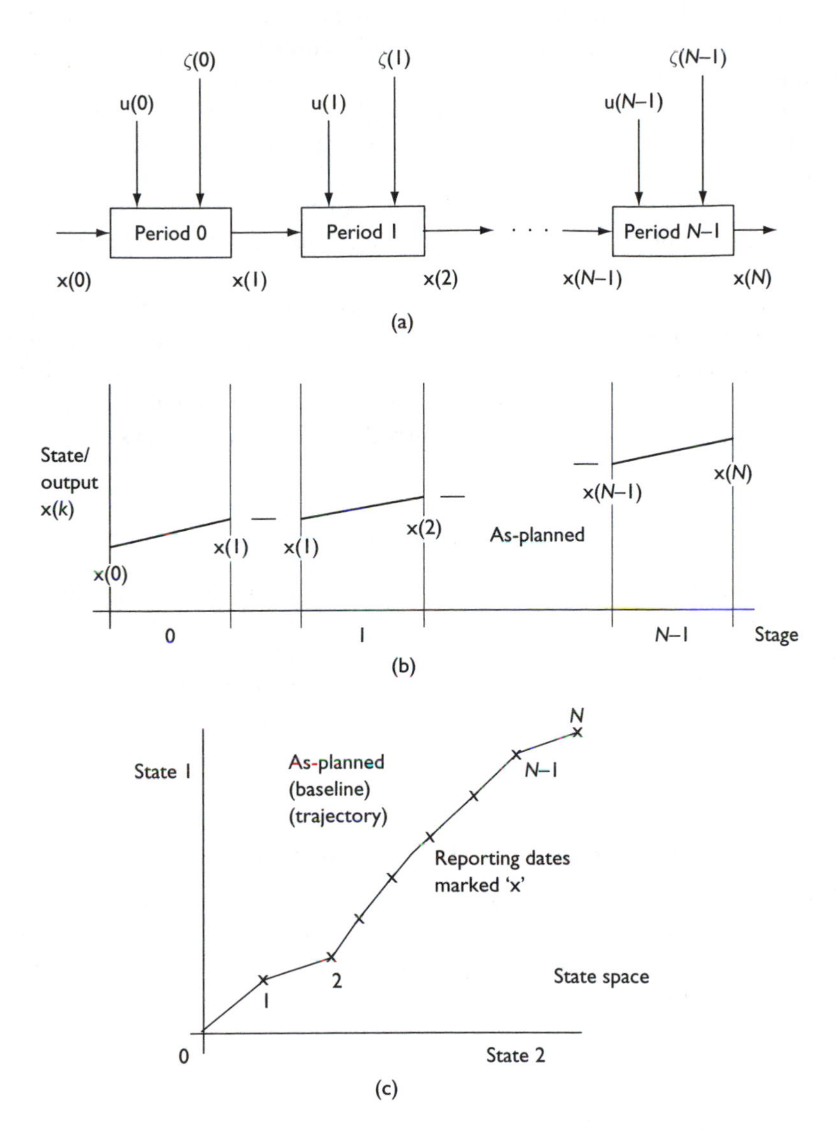

Figure 6.17 (a) Project periods/stages. (b) (Planned) baseline for single state/output. (c) (Planned) baseline for multiple (here 2) states/outputs.

equivalent deterministic influences. (There is no need to talk of risk management.) The control is selected with the view that future system performance is going to be affected by future events/influences. Figure 6.17a shows the project broken up into periods. (The term 'stage' could also be used.) Each period may be any chosen duration and can differ from other periods. At the end of each period, the project performance is monitored and reporting is in terms of absolute performance, and actual performance relative to planned performance. Figure 6.17b shows the (planned) baseline for the case of a single state/output $x(k), k = 0, 1, \ldots, N$. Such a representation is only suitable for the single state/output case. For the case of multiple states/outputs, a state space representation such as Figure 6.17c is necessary. Figure 6.17c shows the (planned) baseline for the example situation of two states/outputs, but it can be extended to higher dimensional state spaces for multiple states/outputs.

Where there is a *deliberate scope change* part way through a project, this will of course lead to a new baseline from the time of introduction of the change onwards. The procedure for obtaining this new baseline is the same as for the original baseline, except that the start time for the optimal control problem is now the time of introduction of the changes.

Project performance better than planned, or as-planned

Should the project performance turn out to be better than planned, or as-planned, then no replanning is necessary, though the planner may choose to take advantage of any good performance, by replanning the remainder of the project. As well, if new or better information comes to hand on future influences, this can be used to refine the baseline for the remainder of the project.

Project performance different to that planned

Replanning options are increased if a change in scope is allowed for the remainder of the project. Allowing for a change in scope is equivalent to allowing a change in the project (system) – a change in the box in Figure 2.5. This in turn mirrors *adaptive* or *learning system* behaviour of control systems theory.

Whether or not a change in scope is allowed, the original baseline can serve as a target at every replanning point. Allowing a change in scope enables adjustment of behaviour to match the baseline when needed. (If no change in scope is allowed, the baseline (target) will change as the project progresses.) Figure 6.18 shows the original (planned) baseline indicated by $x(k), k = 0, 1, \ldots, N$, and the actual behaviour indicated by $xA(k), k = 1, 2, \ldots, N$. The actual behaviour $xA(k)$ will be different to the planned behaviour $x(k)$ because of disturbance $\zeta(k-1)$. The variance is the difference between planned and actual,

$$\mathrm{var}(k) = x(k) - xA(k) \quad k = 1, 2, \ldots, N$$

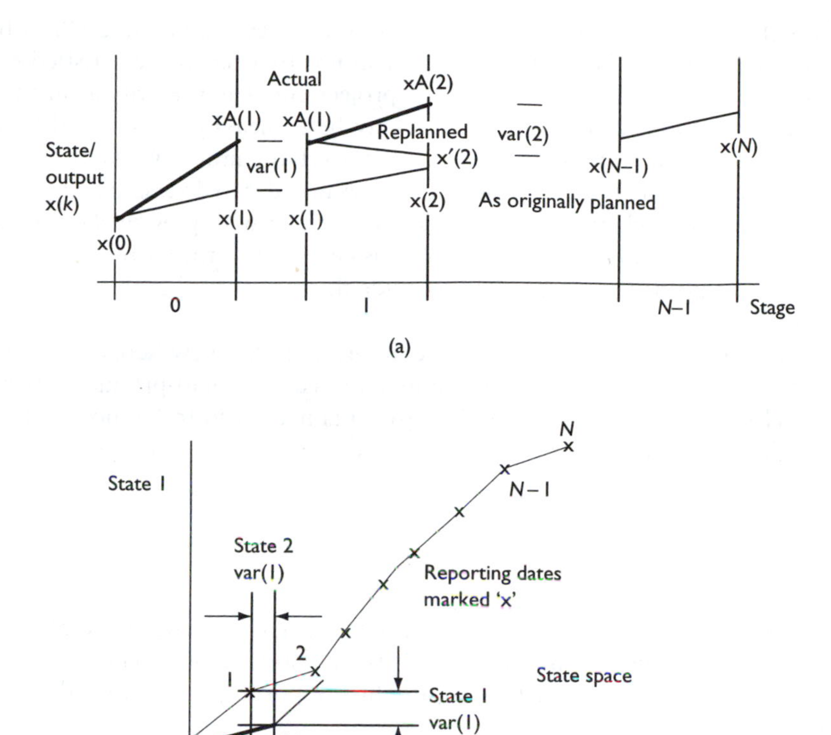

Figure 6.18 (a) (Planned and replanned) baseline and actual project performance for single state/output. (b) (Planned and replanned) baseline and actual project performance for multiple (here 2) states/outputs.

Or if no change in scope is possible, the actual behaviour $xA(k)$ will be different to a revised (updated) planned behaviour $x'(k)$ because of disturbance $\zeta(k-1)$. The variance is the difference between revised planned and actual,

$$\mathrm{var}(k) = x'(k) - xA(k) \quad k = 1, 2, \ldots, N$$

$xA(k)$ can be regarded as the initial conditions for (re)planning the remainder of the project. Commonly, the scope for this remaining part of the project will be different to that envisaged in the initial planning. Solving the deterministic optimal control problem from the current period to the project end (or the equivalent iterative analysis version) will indicate what scope has to be changed (if allowed) and what controls have to be adjusted.

(If no change in scope is possible, this will give a revised baseline $x'(k)$.) If the original baseline is regarded as a constraint that has to be satisfied or approached (or the final state – at the project end – is prescribed) in this optimal control problem, the problem may be so heavily constrained that little optimisation is possible. As before, uncertain future events are converted into equivalent deterministic influences in order to do this calculation. Clearly, solving such an inverse problem is difficult with present-day tools; an iterative analysis-based mode of attack is necessary as the more pragmatic approach using error control ideas to guide the selection of controls for the iterations.

Allowing a change in scope would imply that the new scope may be suboptimal with respect to the original project and end-product objectives. However, the objectives and constraints may need to be modified in the light of project information that has come to hand since the project start.

Example

Consider a project where the planning outcome gave a state space plot as in Figure 6.19. The line in Figure 6.19 is the baseline plot. The numbers $1, 2, \ldots, 10$ represent the 10 dates on which reporting is anticipated to be carried out.

Performance of the project matches the as-planned performance up to reporting date 3, but between reporting dates 3 and 4, actual performance has differed from that planned because of various issues on the project. The state at the latest reporting date is given by 4A in Figure 6.20. As-planned and actual performance are given in Figure 6.20 along with the variances in the states at the latest reporting date.

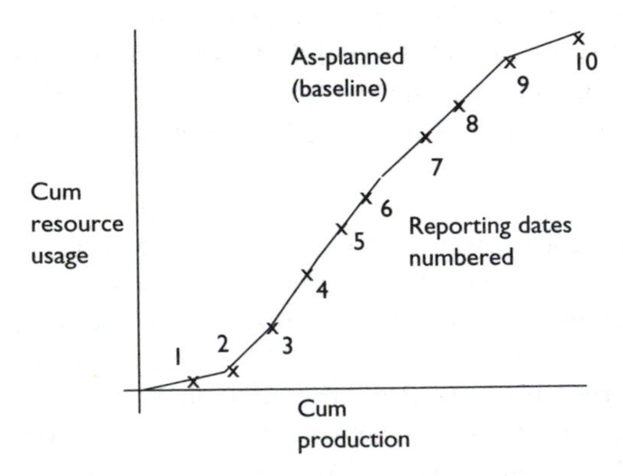

Figure 6.19 As-planned performance (baseline); example.

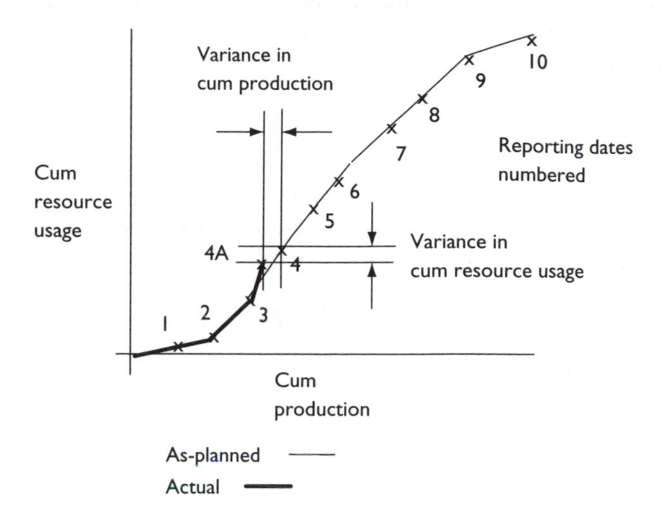

Figure 6.20 Performance, actual and as-planned; example.

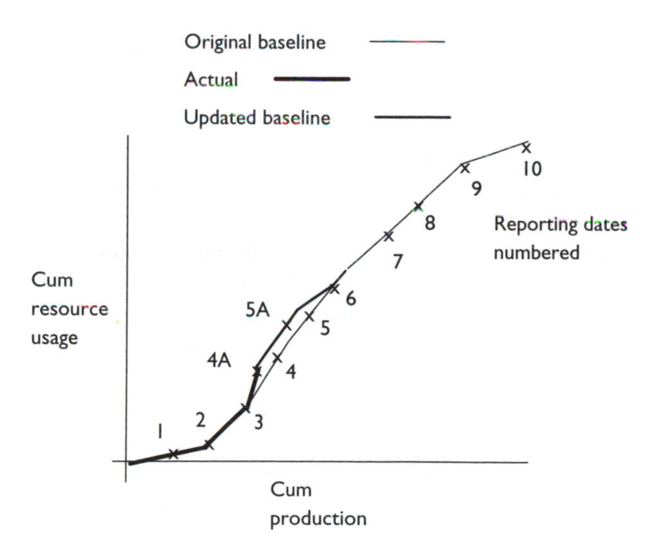

Figure 6.21 Updated baseline; example.

Replanning involves solving the planning problem with initial conditions 4A. The objectives and constraints may remain the same as for the planning problem or may have to be updated to reflect the latest thinking and issues involving the project. Replanning gives the new baseline indicated in Figure 6.21; in this case, the replanning gets the project back onto the original baseline by reporting date 6.

The final state of the project may change.

Exercises

1. How do you establish the best frequency of monitoring a project or activities? Do you monitor at the completion of each activity? Do you monitor after a certain time period? Other?

2. Review the information requested in any project forms to which you have access. Comment on the design of the forms and the need for more/less/alternative information.

3. List reporting problems, other than those mentioned here, of which you are aware.

4. An owner, at pre-tender, estimates a contract package at $2.1 m. The lowest bid from tenderers is $1.8 m and this bid is accepted by the owner. Does the owner's project manager administer the contract based on a budget of $2.1 m or $1.8 m? Why?

5. It is tempting to compare actual costs with the budget. Consider Figure E6.1, where the figure shows the actual cost running above the budget at the latest reporting or status date.

What can be deduced from Figure 6.1 regarding:

- How much is the project over- or under-spending compared with its budget?
- How much is the project ahead or behind production/schedule?

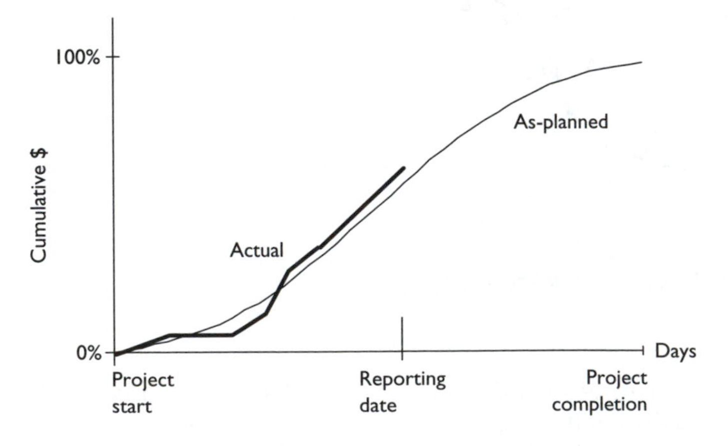

Figure E6.1 Example S curve.

6. Give your views on the best way of obtaining a measure of the per cent complete at any point in a project. You do not have to restrict your thinking to the rules presented above.

7. In earned value reporting, the production/schedule variance is in a unit of money. (This may seem strange.) Why do you think this is so? How might you convert production/schedule variance to a unit of time?

8. What value is there for a cumulative delay plot that is additional to a bar chart showing as-planned and actual? Or, is it redundant information? Does it give management new information?

9. The terms 'variance control' and 'configuration management' are sometimes used to describe the process of watching the scope (of a project). Why do some projects have a need to watch their scope? Do some contractors deliberately play for variations?

10. Given that the early players in a project have the most ability to influence the cost of a project, why is it then that examples continually occur of projects proceeding without thorough preparatory work?

Is there a false economy here of project owners saving money initially in the belief that the total project cost will consequently be less?

(This is reminiscent of the cartoon seen on notice boards in many engineering offices where two 'planners' are looking at the drawings for the Tower of Pisa in Italy [before it was built] and commenting that they could save many Lire if they did not carry out soil tests. The building is now the Leaning Tower of Pisa.)

11. What matters have to be taken into consideration in monitoring, reporting and replanning with respect to cost? What do you regard as the most important matter? What is commitment costing?

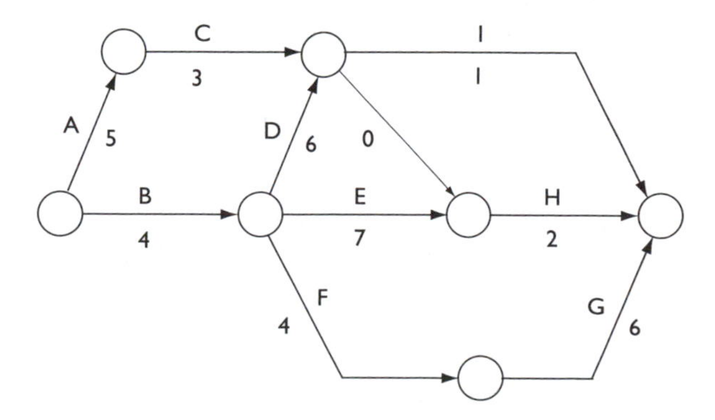

Figure E6.2 Example project network.

Table E6.1 Example project data

STATUS REPORT
PROJECT ....
DAY End of day 10

Activity	*Completed activities*	*Activities in progress*		*Comments*
	Duration (days)	*Days in progress*	*Estimate of days to complete*	
A	6			Material supply problem. Started day 0
B	4			Started day 0
C		2	1	Started day 8
D		4	6	Bad weather delay Started day 6
E		6	1	Started day 4
F	4			Started day 4
G		2	4	Started day 8
H			(2)	–
I			(2)	Revised estimate

12. The status report at the end of day 10 for a project (network shown in Figure E6.2) is given in Table E6.1.

Draw a bar chart showing the original schedule and the activities as-executed. Assume that the original duration estimates of the activities remain unchanged, except where noted in Table E6.1.

Update the network calculations. Draw an updated bar chart for the remainder of the project after day 10.

What remedial measures are possible to get the project back on production/schedule?

13. For monitoring, reporting and replanning practices of which you are aware, identify their strengths and weaknesses.

14. If a scope change occurs, is it correct to regard it as being incorporated into the project? Or, does it create a new project (one with a re-defined baseline)?

15. Comment on the suitability or usefulness of the following contractor's report, as a feed into replanning. Note that a contractor's project report is more selective than an owner's project report.

Project . . . . *Date* . . . . . . / . . . . . . / . . . . . .

Project revenue

Original contract value		$ . . . . . .
Plus variations	Approved	$ . . . . . .
	Pending	$ . . . . . .
	To be submitted	$ . . . . . .
Rise and fall – projected		$ . . . . . .
Projected revenue		$ . . . . . .

Margin calculation

Total project revenue	$ . . . . . .
Less project cost	$ . . . . . .
Contractor's share of savings	$ . . . . . .
Projected margin	$ . . . . . .

Forecast final margin

Best case $ . . . . . .
Worst case $ . . . . . .

Margin trend

Original $ . . . . . .
Previous two reports $ . . . . . . $ . . . . . .
This report $ . . . . . .

Cash trend (POS/NEG)

Previous two reports $ . . . . . . $ . . . . . .
This report $ . . . . . .

Program status

Original completion date	. . . . . . / . . . . . . / . . . . . .
Approved time period extension (days)	. . . . days
Adjusted completion date	. . . . . . / . . . . . . / . . . . . .
Current deviation (ahead/behind) adjusted date	. . . . days
Submitted and likely but unapproved extensions of time period	. . . . days
Extensions of time period not yet submitted	. . . . days
Target completion date	. . . . . . / . . . . . . / . . . . . .

Turnover

Turnover achieved this month	$ . . . . . .
Cumulative turnover to date	$ . . . . . .
Forecast turnover to end of project	
Next month	$ . . . . . .
Month 2	$ . . . . . .
Month 3 etc.	$ . . . . . .
Total	$ . . . . . .

Chapter 7

Selected topics

People issues

Introduction

The development in this book largely centres on the mechanics of planning. The place of people in planning has been deliberately excised from this, in order to present a clarity of approach. But of course people are involved in planning, and whenever people get involved in anything, that thing takes on inexplicable complications; what in principle may be straightforward takes on new dimensions of complexity.

Human resource or people management is a very imprecise art, with most literature on the topic being subjective, opinion and anecdote based; it lacks objectivity and an ability to justify through the scientific method. It also tends to be faddish because of this. The following notes are no different in that alternative views can be expressed with equal enthusiasm.

Safety, industrial relations and quality involve people issues, in particular, that influence planning. An organisation's policies (general guides for decision making and individual actions), procedures (detailed methods for carrying out policies) and standards (degree of individual or group performance defined as adequate or acceptable) also influence planning practices.

Planning is part of a project manager's responsibilities, but involves all personnel in a project to some degree. In some cases, the responsibility for preparing the main project plan is that of a program manager assisted by a program team, under the overall direction of the project manager.

There is a project role involving coordination and the interrelationships of activities and project participants, and in providing continuity of work for the project participants. When scheduling people as a resource, this may be done only as a group, team or gang, rather than as individuals.

Planning well done and realistic can be motivational to the project team.

Involvement of people

Planning for projects ideally involves contributions from everyone who will be involved in the implementation of the plan, although there may be organisational difficulties with achieving this. In practice, there are people who do the planning (the planners) and there are people – end-users or doers – who implement the plans, yet it is believed that there is merit in the latter group contributing their expertise to the planning and in being consulted in the development of any plans.

It is sometimes said that the success of a plan is directly proportional to the participation of those responsible for carrying out the planned work. Commitment to the plan comes from an involvement in developing the plan, and gaining some ownership of the plan. There is even a case in some circumstances of having decentralised planning; that is, the people responsible for the performance of the project do the planning. Generally, a plan should not be imposed on others; agreed realistic plans are more likely to be successful. There can be a view of doers, uninvited to participate in the planning, that the plan represents achieving the impossible.

The involvement of end-users in the planning can have additional productivity and quality-related benefits such as:

- Giving to them a better understanding of their job
- Giving an indication of where they fit into the overall scheme of the project
- They gain confidence in the presumably more efficient and more effective work methods; they can assume that they are working smarter not harder
- The reasons for proper reporting, feeding into replanning, become more readily understood
- Reporting is seen to be for the project's benefit rather than 'big brother' checking up on employee performance.

However, the planning process can be tainted from involving the end-users in the planning because of their lack of appreciation of planning and the concentration on their own details at the expense of the bigger picture. There have also been many successful projects where the end-users had no involvement in the planning. This leads to the question of how much involvement of the end-users should there be in planning?

Where people have to accept the work of planners, without having an involvement, the planners have to be realistic and credible, and take into account the views of others. Planners may be part of a larger service department within an organisation (Figure 7.1). This may create a them-and-us attitude unless the planners are seen as genuine members of the project team.

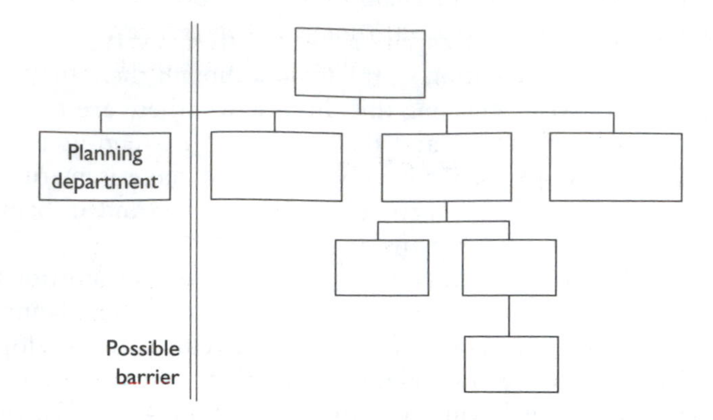

Figure 7.1 Possible barrier between the 'planners' and the 'end-users'.

Project personnel should be aware of planning principles and good planning practice, though they may only exercise a contributing role and may not be directly involved in the mechanics of planning. Planning is a discipline that leads to informed management.

As well, top or senior management commitment and support for a project, and the awareness of this support by project personnel, is believed to be a necessary ingredient of a successful project. It follows that senior management must provide the necessary resources in order to undertake the planning.

The above refers to the structured planning carried out on projects. But everyone on a project will carry out some modest, possibly unstructured, planning of their own.

Work breakdown and initial planning

In initial planning, there may be a need to bring together various parties in order to provide their specialist input, or to provide alternative views.

In developing a work breakdown there is a general feeling that people are only familiar with one or two levels below the level of their position; beyond that it is too remote and they need assistance from those contributing at lower levels. (There are some people, though, who have risen through organisational levels and who may be capable of going further.) Input from people who are responsible for the work (and later monitoring the work) may be necessary. This aligns the requirements of the end-users and the planners.

Good practice sees that the work breakdown structure (WBS) is reviewed by all people with an involvement in the work, and revised as necessary, ensuring that any revisions are also reviewed.

Barriers to effective planning

Many of the barriers to effective planning are people related and people attitude related. Improvements in planning can be gained by recognising the existence of these barriers. Some barriers are:

- Many people find thorough planning tedious, and would prefer to start 'doing' before giving planning enough consideration.
- Many people feel confident relying on their experience and judgement to handle troubles when they arise ('crisis management', 'shooting from the hip'), rather than working through planning in a thorough way and making prior allowances for possible eventualities. People falsely believe that thorough planning stifles initiative.
- Many people resent the time extent and cost involved in thorough planning, and do not connect project success with thorough planning. Other work pressures and poor personal 'time' management skills mean that little effort is devoted to planning.
- Planning involves creativity and the generation of alternatives and ideas. Many people do not have such skills. People constrain themselves to doing something today the way it was done yesterday. Lateral thinking is not undertaken, and in some industries is discouraged.
- People have trouble converting the owner's value system into workable objectives and constraints. The development of project objectives and constraints is commonly poorly done. And everything in planning depends on what the objectives and constraints are.
- Senior management often has poor attitudes to thorough planning and its cost effectiveness. This is reflected in poor support/resourcing of the planning activity.

Responsibility matrices

For each activity, there are attendant people resource requirements. People resource and organisational (including coordination) requirements, responsibilities and authorities can be represented in a responsibility matrix/chart form which may have rows corresponding to the activities, columns corresponding to people resources and entries corresponding to responsibilities, or similar (Figures 7.2, 7.3 and 7.4).

The matrices are also a useful way of communicating these responsibilities, roles and duties to people involved in the project. Such charts show where different project personnel interface and where managerial coordination is required. They relate activities to people, to groups of people, to departments and so on.

It is important to note that everyone knows who is doing what, that is, that they know the division of responsibility. Without a clear idea of

		Responsibility					
		Project manager	Project engineer	Planning engineer	Office engineers	Field engineers	. . .
Subproject	Activity						
A	A1						
	A2						
	A3						
	A4						
	A5						
	A6						
B	B1						
	B2						
	B3						
	B4						
	⋮						

Legend for entries:
Δ Responsibility
s Support
◊ Notification
a Approval
c Consultation
v Supervision

Figure 7.2 Example responsibility matrix/chart.

responsibility and activity assignment, people can become confused. People double up on some jobs, while some jobs do not get done at all. Project performance suffers.

Replanning issues

When replanning, besides the usual controls related to work method, resources and resource production rates, there are a number of people-related things that a project manager can try to improve a project's performance, such as:

- Improved leadership and management
- Improved coordination of workers and equipment.

	Person 1	Person 2	Person 3	Person 4	Person 5	Person 6	etc.
Activity A	5		2	4			
Activity B		3				5	
Activity C					6		
Activity D			1		1		
etc.							

Key
1 Actual responsibility
2 General supervision
3 Must be consulted
4 May be consulted
5 Must be notified
6 Final approval

Figure 7.3 Example responsibility matrix/chart.

Activity	Responsibility		
	Initiator	Coresponsibility	Approval
A1			
A2	Enter title or person's name		
A3			
A4			
. . .			

Figure 7.4 Example responsibility allocation.

Associated with monitoring and reporting may be regular project review meetings.

It can be useful to tie monitoring in with the organisational structure of the project so that employees do not perceive the monitoring as something external and imposed upon them, but as an essential part of their work.

Data collection is carried out by project staff who generally may rather be doing something more tangible. Enthusiasm for filling in forms is not there and so the need for the data has to be insisted on. Feedback in a form noticeable to those collecting the data assists by demonstrating the usefulness of the data.

It is recommended that regular reporting be part of the statement of duties of project personnel, and all personnel connected with the project be tied into the reporting system.

Honesty in reporting the true status is important for effective replanning. The human frailty of embellishing the truth, in order that it reflects better on the person responsible, has to be avoided.

Learning curves

Gaining the benefits from repetitive work can only happen with people involvement. Increased experience permits the improvement.

Case example – Computerised catalogues

Reporting and human behaviour are the focus of this case example. These are discussed with respect to a start-up software company that specialised in the development of computerised component catalogues.

The company produced computerised component catalogues for discrete part manufacturers. The company had eight employees – two managers and six programmers.

Because of the small number of employees, very few formal procedures were ever implemented. Work orders and reports were communicated verbally, with no written records. There were no formal project plans and work progressed according to the needs of the day, as defined by the managers. The result was numerous late deliveries of computerised catalogues, and the lateness only became apparent as the deadline approached. There was also no data from which to learn lessons and plan for similar projects, and consequently mistakes were often repeated, not only by different programmers but also by the same programmers.

An example of this related to the building of the setup program to install a computerised catalogue. This task is a standard one and only requires minor modifications for each catalogue (for example, company file, file paths and names). For a particular hydraulic component catalogue, a problem was found regarding the manner in which the setup program determined which system files to install and which not to install. The routine used by the catalogue install program was causing some of the client's other programs to cease to work after the catalogue's installation. This situation was obviously very embarrassing to the company and a workaround was found and implemented as quickly as possible.

The next catalogue that was completed, in this case a vacuum component catalogue, was prepared using the older version of the installation program template and hence was sent to the client without the new workaround. This mistake was only realised when the client complained that several programs were locking up after installation of the catalogue. The failure to properly report and learn from this problem made the company look very unprofessional to important clients. It also caused much frustration to the programmers who took pride in their work.

More recently, a weekly production meeting was instituted, mainly for the purpose of reporting in both directions. A summary of the latest meeting was kept and progress could be determined from this. This did not replace the informal, verbal communications but rather added a formal element that was sorely missing, as well as a historical record. In this way, the company could manage schedule and quality matters relating to the development of computerised catalogues.

The company programmers were all middle aged, and were sensitive to the issue of supervision. In one weekly production meeting, when presented with the idea of keeping formal timesheets, one employee compared management to an authoritarian government. Monitoring and reporting were looked upon with suspicion and the whole issue had to be handled with care.

In spite of this attitude, a monthly timesheet was adopted for monitoring the cost of the catalogue development projects. This was chosen as the method because it was not very intrusive and the worker maintained influence over this task. The issue of dishonesty was not considered to be great because of the personalities of the employees and because, as the number of employees was small, management was in constant contact with everyone. As the company grows, however, it is understood that a more rigid system will have to be implemented, such as an automatic schedule tracking program residing on the network that can 'see' what programs and projects are being run on all the computers. How the employees respond to such a system is yet to be seen. It is foreseeable that some employees may opt to leave rather than work under such supervision.

The company situation with respect to reporting and the employees' reaction to both reporting and monitoring is very sensitive. Unfortunately, the need for proper reporting and replanning in software projects is critical in order to achieve a successful project. This conflict creates a difficult problem for the project manager.

Exercises

1. Where do the monitoring and reporting personnel fit in the organisational structure of a project team?

Can the monitoring and reporting be done by the planners?

What communication is necessary between the monitoring and reporting personnel and the planners in order that the project plan is continually kept up to date?

2. Suggest ways of making monitoring, reporting and replanning an accepted part of the work environment.

3. Planning calculations indicate which are the critical activities. Following disturbance, replanning may indicate changed critical activities; some non-critical activities may become critical, while some critical activities may become non-critical. How do you stop people from overreacting to such changes?

4. If a project person is trying to hide something that is his/her fault, how do you get accurately reported information on a project's performance, especially with a large span of management? Should the people (namely the planning staff) who are going to use the project data be the ones who collect the data? This would have an added benefit of avoiding communication errors. What role could the equivalent of an independent third party or auditor have in data collection and reporting?

Monte Carlo simulation

Introduction

Ordinarily, planning is converted from a true probabilistic problem to a deterministic problem for tractability reasons (Figure 7.5). However, it is possible to acknowledge the probabilistic nature of projects.

Project activities generally have variability associated with them, leading to uncertainty in the planning formulation. The variability in the estimate

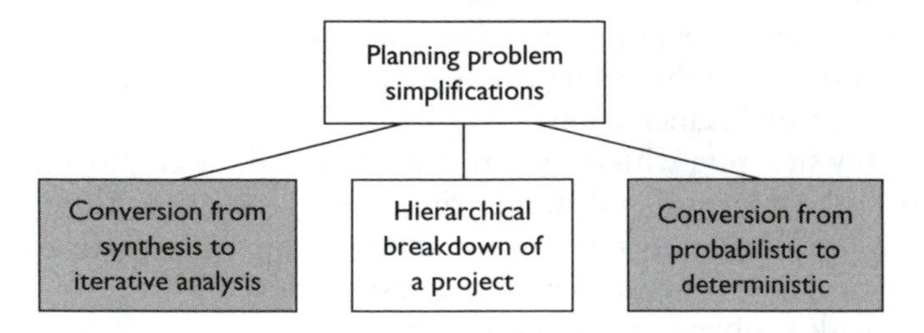

Figure 7.5 Simplifications to make the planning problem more tractable.

of an activity's duration may arise from the work itself or from the estimator having insufficient data to characterise the work exactly. This has led to the consideration of techniques such as *PERT* (program evaluation and review technique), and *Monte Carlo simulation* in the planning of projects (in an iterative analysis form). These notes examine Monte Carlo simulation. Monte Carlo simulation offers a practical way to incorporate variability. The planning of a project should ideally take into consideration the variability in the durations of the project's component activities and the consequent varying resource demands, but usually this is not done in practice.

Monte Carlo simulation is a technique for analysing nearly anything that contains probabilities, and not just project networks.

PERT, in various forms, has been around since about the 1960s, with seminal credit usually being given to navy research and development work. PERT introduces uncertainty into activity duration estimates. Planners are asked to estimate activity durations based on optimistic, pessimistic and most likely considerations.

Calculations are performed on an expected activity duration together with an associated measure of the variability of this duration. It is a first-order method. It involves approximations.

The main information coming from a PERT-type analysis is the probability of completing a project on a particular date. Consequently, attention has focused on Monte Carlo simulation as a more general stochastic technique to provide additional information such as:

- The probability of completing a project at a given cost
- The probability that an activity will become critical or not
- How to increase the probability of project completion by a specific date by increasing resource numbers to some activities in the network
- Probabilistic cost diagrams or cash flow diagrams which help the decision maker.

Monte Carlo simulation in the analysis of networks appeared in the 1960s. The technique involves repeatedly generating sets of random numbers, transforming these in accordance with desired probability distributions of the activity durations, and obtaining samples of the relevant network results. The results are treated statistically with presentations in the form of histograms, and methods of statistical estimation and inference are applied. For these reasons, the Monte Carlo method is also a sampling technique, and as such shares the same problems of sampling theory, namely the results are also subject to sampling errors. Generally, therefore, Monte

Carlo simulation which results from finite samples are 'not exact' (unless the sample size is very large). Monte Carlo simulation provides a significant analysis tool useful in (the iterative analysis form of) planning. Most people agree that the only way to realistically incorporate variability in activity durations for large complicated networks is to use the technique of Monte Carlo simulation. Monte Carlo simulation has the ability to use overlapping relations without creating subactivities.

Monte Carlo simulation, in effect, converts a difficult probabilistic analysis problem into many simpler deterministic analysis problems. The steps involved for networks are:

- Generating activity durations for each activity (via uniform random numbers)
- Carrying out a conventional network analysis (incorporating overlapping relationships, if applicable)
- Repeating the above two steps many times
- Collecting relevant statistics on important items, such as project completion dates.

The first step is one of data generation, the second step is analysis, while the fourth step is bookkeeping.

Monte Carlo simulation is suited ideally to computer use, where large amounts of data can be handled with ease. It is not a technique for use by hand or where a closed-form solution is required.

The term 'variance' in the following refers to the probabilistic/statistical variance.

Activity duration generation

The duration of each activity is described by some probability distribution (Figure 7.6). Sampling from each activity's distribution is accomplished via the use of uniform random numbers.

Uniform random numbers are numbers that have an equal probability of occurrence. The associated probability distribution looks like Figure 7.7. Figure 7.7, in fact, is drawn for the *standard* uniform random variable which takes values only between 0 and 1, and is distributed uniformly between these values.

Drawing the second parts of Figures 7.6 and 7.7 next to each other and to the same scale on the vertical axes gives Figure 7.8.

Without performing any mathematics, it appears reasonable that

$$y_i = F_Y^{-1}(v_i) \quad i = 1, 2, \ldots$$

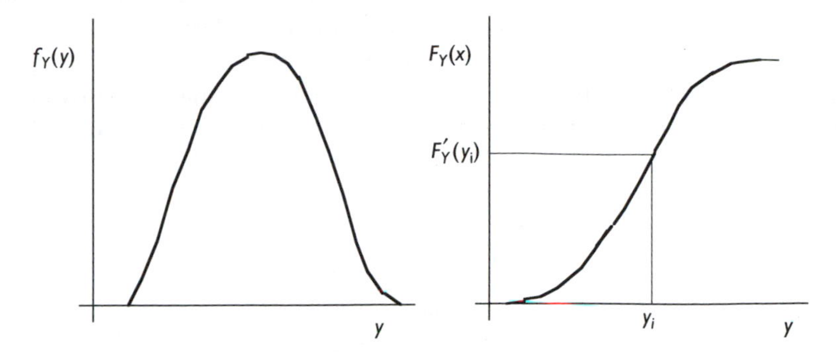

Figure 7.6 Probability density function (PDF) and cumulative distribution function (CDF) for a typical activity's duration.

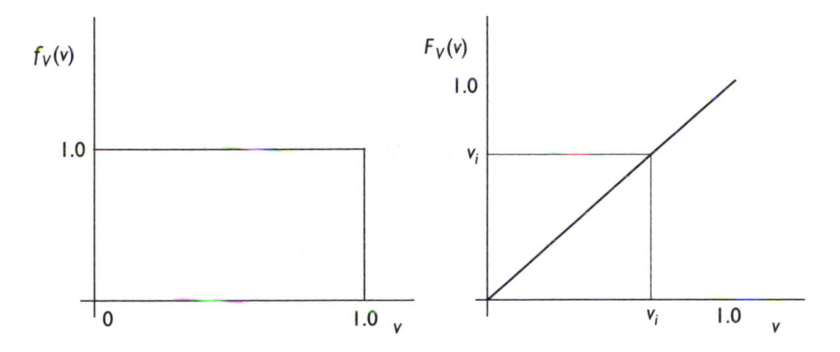

Figure 7.7 PDF and CDF for standard uniform variate V.

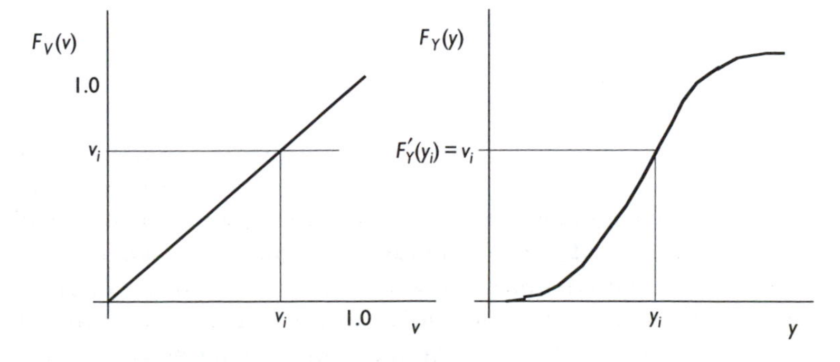

Figure 7.8 Transformation between V and Y.

That is, if $(v_1, v_2, \ldots, v_n)$ is a set of values from V, the corresponding values for y_i, $i = 1, 2, \ldots, n$, are obtained through this last expression. This is sometimes referred to as the *inverse transform method*.

Some distributions allow the inversion of the cumulative distribution function in closed form. For example the exponential distribution, where,

$$F_Y(y) = 1 - e^{-\theta y} = v$$

then

$$e^{-\theta y} = 1 - v$$

or

$$y = \frac{-\ln v}{\theta}$$

Generally, however, the inversion of cumulative distribution functions is not possible in closed form, or it leads to unattractive computations. Other means are then used to transform between the standard uniform variate V and the activity duration Y.

For an Erlang distribution of type ℓ with mean $1/\mu$, the transformation commonly used is:

$$y = -\frac{\ln \prod_{i=1}^{\ell} v_i}{\ell \mu}$$

For a normal (Gaussian) distribution with mean μ and variance σ^2, the transformation commonly used is:

$$y = \sigma \left(\sum_{i=1}^{12} v_i - 6 \right) + \mu$$

Note that to generate a single sample from an Erlang or a normal distribution, more than one standard uniform random number is required.

For discrete random variables and where probability density functions (PDF) are discretised, the transformation is based on the shape of the probability mass function (PMF). For example, consider an activity duration with the probability mass function as shown in Figure 7.9.

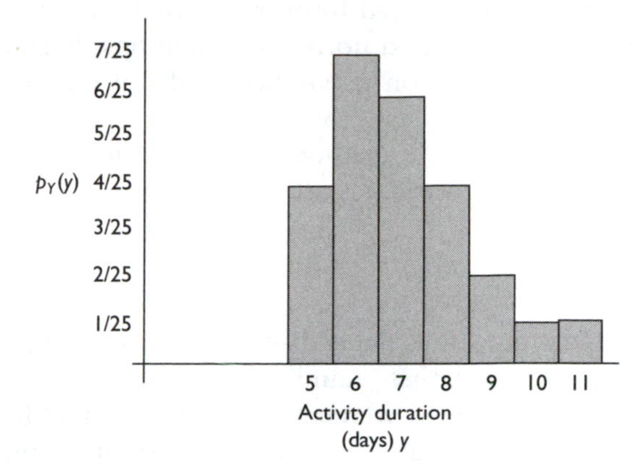

Figure 7.9 Example probability mass function (PMF).

There,

$$P[Y = 5] = 0.16$$
$$P[Y = 6] = 0.28$$
$$P[Y = 7] = 0.24$$
$$P[Y = 8] = 0.16$$
$$P[Y = 9] = 0.08$$
$$P[Y = 10] = 0.04$$
$$P[Y = 11] = 0.04$$

That is, 16% of the area under the probability mass function corresponds to $Y = 5$, 28% to $Y = 6$ and so on. It follows that 16% of the random numbers generated can be allocated to $Y = 5$, 28% to $Y = 6$ and so on. For v_i lying between 0 and 1, it is convenient to adopt a transformation as outlined in Table 7.1.

Table 7.1 Selected relationship between v and y

Value of v_i *generated*	*Corresponding value of* y_i
$0 \le v_i < 0.16$	5
$0.16 \le v_i < 0.44$	6
$0.44 \le v_i < 0.68$	7
$0.68 \le v_i < 0.84$	8
$0.84 \le v_i < 0.92$	9
$0.92 \le v_i < 0.96$	10
$0.96 \le v_i \le 1.0$	11

This is a numerical equivalent to the closed-form transformations exampled earlier for the exponential, Erlang and normal distributions. For any discrete random variable, each distribution is handled in this way, and a unique transformation set up.

In the simulation literature, alternative methods of transformation for both continuous and discrete random variables are given.

Uniform random numbers

The above generation of activity durations depends on there being a supply of uniform random numbers. How are these numbers obtained?

There are numerous mechanical ways by which uniform random numbers can be obtained, for example rolling a dice, tossing a coin or spinning a roulette wheel. For a dice, the numbers 1 to 6; for a coin, the numbers 1 (heads) and 0 (tails); for a roulette wheel, the numbers 0 to 37 (or 38) all have an equal probability of occurrence.

Uniform random numbers are also available tabulated in appendices of many books on statistics. Calculators, spreadsheet packages etc. have inbuilt random number generation facilities as well.

Such means of obtaining uniform random numbers are not convenient, however, if a computer-oriented approach to analysing networks with probabilities is sought.

An alternative way of generating uniform random numbers is to use what are called 'pseudorandom number generators' which are mathematical expressions/numerical algorithms. They produce numbers which are distributed almost uniformly, and for most project purposes any small difference between genuinely uniform and almost uniform is of no consequence.

One example of a pseudorandom number generator is (for example, Ang and Tang, 1984)

$$r_{i+1} = (ar_i + c)(\text{mod } m)$$

which uses modulo arithmetic. The right-hand side translates as the remainder after $(ar_i + c)$ is divided by m. Let

$$k_i = \text{Int}\left(\frac{ar_i + c}{m}\right)$$

Then

$$r_{i+1} = ar_i + c - mk_i$$

In these expressions, r_i, $i = 0, 1, 2, \ldots$, are random numbers; a, c and m are non-negative integers chosen by the user (c may sometimes be chosen as 0).

These random numbers may be normalised by dividing by the modulus m,

$$v_{i+1} = \frac{r_{i+1}}{m} \quad i = 0, 1, 2, \ldots$$

and these constitute a set of random numbers between 0 and 1 with the standard uniform probability distribution.

Numbers so generated are not strictly random because the whole sequence can be duplicated in a deterministic fashion. The numbers are accordingly called ‘pseudorandom’.

The algorithm starts by assuming a seed value, r_0, calculating r_1, then r_2, then r_3 and so on.

For example, let $a = 3, c = 1, m = 7$ and $r_0 = 1$. The calculations proceed as follows:

$$i = 0 \qquad k_0 = \text{Int}\left(\frac{3 \times 1 + 1}{7}\right) = \text{Int}\left(\frac{4}{7}\right) = 0$$

$$r_1 = 3 \times 1 + 1 - 7 \times 0 = 4$$

$$v_1 = 4/7 = 0.571$$

$$i = 1 \qquad k_1 = \text{Int}\left(\frac{3 \times 4 + 1}{7}\right) = \text{Int}\left(\frac{13}{7}\right) = 1$$

$$r_2 = 3 \times 4 + 1 - 7 \times 1 = 6$$

$$v_2 = 6/7 = 0.857$$

and so on.

This leads to the numbers in Table 7.2.

Note that this provides a range of random numbers between 0 and 6, the numbers repeat with a cycle length of 6 and not all numbers between

Table 7.2 Example generation of random numbers

i	r_{i+1}	v_{i+1}
0	4	0.571
1	6	0.857
2	5	0.714
3	2	0.286
4	0	0.0
5	1	0.143
6	4	0.571
7	repeats	

0 and 6 are obtained. The cycle length is related to m and the standardised random numbers are obtained by dividing by m.

For a useable set of random numbers, m has to be large to avoid the possibility of the same numbers recycling. Many texts suggest using, for example, values of m about 2^{30} to 2^{35}. The value of r_0 is usually chosen between 0 and m.

For the case $c = 0$, m should not be a multiple of k, and r_0 should not be chosen as 0, else all random numbers will become 0.

Generally, values of v_i of 0 or 1 should be excluded from the generated set of random numbers.

Example

Figure 7.10 is an elementary network chosen to illustrate all the steps in applying Monte Carlo simulation to a network analysis. An activity-on-node diagram is used although the technique is equally applicable to both network types. The technique is not restricted by network size or complexity.

Each activity in the network has its own unique distribution describing the variability in its duration. It is assumed here that the distributions are of the Erlang type. Activity duration data are given in Table 7.3.

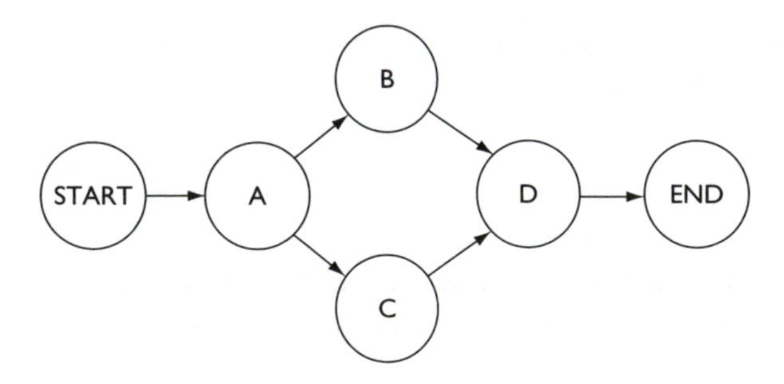

Figure 7.10 Example network.

Table 7.3 Example network data

Activity	*Duration distribution*	
	Erlang shape parameter	*Mean (days)*
A	5	5
B	5	7
C	5	8
D	5	3

Step 1

The process starts by generating sufficient random numbers in order to sample from the activity distributions. Since the four activities are each modelled by Erlang distributions with shape parameters of 5, then $4 \times 5 = 20$ random numbers are required.

Using one particular pseudorandom number generator, for example, the first 20 random numbers (standardised to between 0 and 1) turn out to be:

0.0006, 0.0011, 0.1855, 0.1536, 0.6107, 0.3812, 0.4812, 0.1803, 0.0647, 0.0995, 0.6912, 0.7502, 0.7533, 0.8062, 0.7061, 0.0040, 0.0672, 0.1422, 0.4166, 0.0822.

Step 2

These 20 random numbers translate into 4 samples of activity durations. For example, for activity A

$$\begin{aligned} t_A &= -\frac{5}{5} \ln(0.0006 \times 0.0011 \times 0.1855 \times 0.1536 \times 0.6107) \\ &= 18.28 \text{ days} \end{aligned}$$

By the same process the sampled durations for activities B, C and D are respectively,

$$t_B = 11.84 \text{ days}; \quad t_C = 2.41 \text{ days}; \quad t_D = 8.13 \text{ days}$$

Step 3

A conventional critical path analysis of the network is carried out using the activity durations calculated in Step 2 (Table 7.4). (This analysis may incorporate overlapping relationships if applicable.)

This information is collected and summarised with similar information at the end of the simulation.

The simulation now repeats Steps 1, 2 and 3 many times.

Table 7.4 Results of critical path analysis on example network, first cycle

Activity	*Dur. (days)*	*EST*	*EFT*	*LST*	*LFT*	*TF*	*FF*	*IF*	*Critical*
A	18.28	0.0	18.28	0.0	18.28	0	0	0	*
B	11.84	18.28	30.12	18.28	30.12	0	0	0	*
C	2.41	18.28	20.69	27.71	30.12	9.43	9.43	0	
D	8.13	30.12	38.25	30.12	38.25	0	0	0	*

Step 1

Step 1 is repeated. That is, a further 20 random numbers are generated. Using the same pseudorandom number generator, these are:

0.3969, 0.7478, 0.7133, 0.1255, 0.1332, 0.2639, 0.4867, 0.2745, 0.6665, 0.3310, 0.6276, 0.6699, 0.3876, 0.5898, 0.0266, 0.4522, 0.6871, 0.6808, 0.5742, 0.7621.

Step 2

Step 2 is repeated giving

$$t_A = 5.64 \text{ days}; \quad t_B = 6.80 \text{ days}; \quad t_C = 13.21 \text{ days}; \quad t_D = 1.43 \text{ days}$$

Step 3

The network is analysed using the latest sampled activity durations (Table 7.5).

Note that the critical path has changed. The path now runs through activity C rather than activity B. As a general comment for all networks, the critical path(s) may change from one analysis to the next throughout the simulation.

The process again returns to Step 1.

Final step

Steps 1, 2 and 3 are repeated many times. Each time Step 3 is carried out information on activity start times, activity finish times and activity floats is collected. The final step summarises this collected information. More information implies greater reliance can be placed on the results and so it is not uncommon for Steps 1, 2 and 3 to be repeated 1000, 10 000 or more times. It is tedious work but work at which a computer excels. Computer run durations are obviously longer than in the deterministic case but even with large networks, the computer run duration is not a concern.

Table 7.5 Results of critical path analysis on example network, second cycle

Activity	*Dur. (days)*	*EST*	*EFT*	*LST*	*LFT*	*TF*	*FF*	*IF*	*Critical*
A	5.64	0.0	5.64	0.0	5.64	0.0	0.0	0	*
B	6.80	5.64	12.44	12.05	18.85	6.41	6.41	0	
C	13.21	5.64	18.85	5.64	18.85	0.0	0.0	0	*
D	1.43	18.85	20.28	18.85	20.28	0.0	0.0	0	*

Commonly the collected information is summarised in histogram form. For example consider information collected on the project completion times (the finish times of activity D). After α passes ($\alpha = 1000$, 10 000, etc.) through Steps 1, 2 and 3, there are α values of project completion times. By usual means these values are converted to a histogram such as Figure 7.11. Generally as α gets larger, it could be expected that the shape of the density function for the project completion time approaches more closely the shape of the density function for the project completion time had the network been analysed in a closed-form mathematical fashion. It is generally acknowledged though that a closed-form analysis is difficult for all but the most elementary networks with elementary assumptions on the probability distributions describing the activity durations.

From Figure 7.11 it is possible to evaluate details such as the probability that the project will be completed by a certain date. Probabilities of completion by a certain date are calculated from the area of the histogram to the left of that date; probabilities of exceeding a certain date are calculated from the area of the histogram to the right of that date.

For projects containing many activities, this histogram could be expected to approach a normal distribution in shape according to the central limit theorem. This is more so for a cost histogram because all activities contribute to the total cost.

Criticality index

Monte Carlo simulation allows the calculation of a criticality index (Van Slyke, 1963). The criticality index is the probability that an activity will be on the critical path. The criticality index is defined as $W(a)/M$, where

$$W(a) = \sum_{i=1}^{M} w_i(a)$$

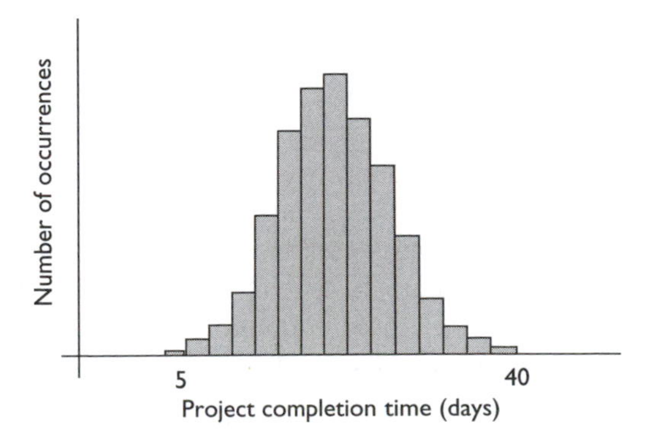

Figure 7.11 Example histogram of project completion times.

and

$$w_i(a) = \begin{cases} 1 & \text{if activity } a \text{ is critical during simulation experiment } i \\ 0 & \text{otherwise} \end{cases}$$

for M simulations.

For example, if an activity is on a critical path 3000 times out of 10 000 simulation runs, this activity is given a criticality index of 0.30.

Monte Carlo simulation can accommodate three options to establish the probability of any activity becoming critical, that is the criticality index:

(i) The latest finish date of the project is set equal to its earliest finish date. In this case, the non-critical activities in the network have positive total floats and every network analysis contains critical activities.
(ii) A planned finish date is assumed and the latest finish date of the project is set equal to that date. In this case, some activities may have a negative total float and in some simulations the network contains no critical activities.
(iii) This option is similar to the second option, the difference being that the planned project finish date is not given but rather deduced from a given number of simulations.

Option (i), in general, gives a higher criticality index. This follows from the observation that a critical path occurs in every simulation. This is to be compared with Options (ii) and (iii) where critical paths do not occur in every simulation.

In Option (i), the Start and End activities have criticality indices of 100%; that is, these activities have a probability of 1 of becoming critical. The Start and End activities in Options (ii) and (iii) usually have probabilities less than 1 of becoming critical; this value, however, is higher than probabilities associated with other activities becoming critical.

Option (i) leads to positive total floats, while Options (ii) and (iii) may give positive and negative total floats.

Cost inclusion

There are a number of ways by which costs can be included in a Monte Carlo simulation. The approach given here is that described in Al-Sadek and Carmichael (1992).

Each activity is assigned a mean value of cost, C_m, and a ratio, R, of labour and equipment cost to material cost. For example if $R = 0.1$, this means that the cost of the activity mainly consists of material, while if $R = 0.9$ the cost of the activity mainly consists of labour and equipment.

It is further assumed that the cost of an activity is proportionally related to the activity duration at a rate equal to R. This is based on the assumption that the cost of materials stays constant but that the cost of labour and materials varies as the activity duration varies. Also, it is assumed that the resource numbers used for any activity do not change even though the activity duration changes. That is,

$$C_a = C_m + \frac{(T_a - T_m)}{T_m} R C_m$$

where

C_a is the cost of the activity at a given duration T_a
T_m is the mean activity duration (Figure 7.12).

For each activity, a duration T_a is generated from its associated probability density function. This also establishes C_a from the above equation. Usual forward and backward passes are then performed on the network with earliest times, latest times, floats and costs being evaluated for each activity.

Example

Consider the construction of a three-span bridge with precast concrete beams and piled foundations (based on Carmichael, 1989). Table 7.6 lists the component activities, their mean estimated durations (T_m), their shape parameters for the assumed probability (Erlang) distribution, the ratios of labour and equipment to materials cost (R) and the mean values of cost (C_m). The time unit is day; costs are expressed in a suitable money unit. The activity-on-node diagram for this project is given in Figure 7.13. The translation of the activity codes used in Figure 7.13 is given in Table 7.6.

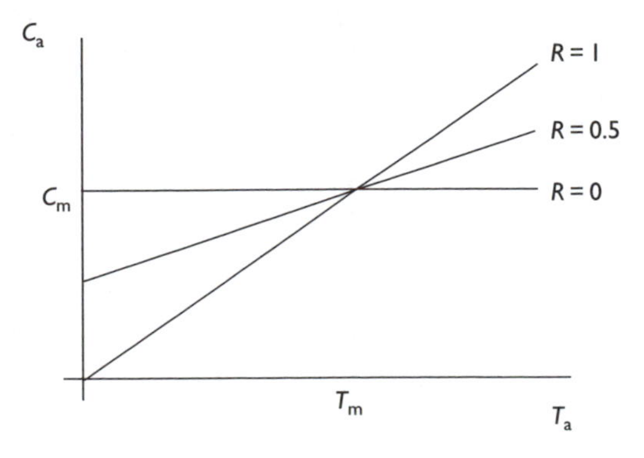

Figure 7.12 Relationship between the cost and the duration of an activity.

Table 7.6 Three-span bridge data. (Activity names have been abbreviated because of space limitations – work items are implied.)

Activity code	*Activity*	T_m	*Erlang Shape Parameter*	R	C_m
	Establishment activities				
Start	Start up	1	10	0.8	1000
Detour	Set up road detour	1	10	0.5	14 500
Site Prep.	Site preparation	5	5	0.5	10 000
	Order and delivery activities				
O/d Piles	Piles	40	20	0.2	61 000
O/d Rein.	Reinforcement	5	20	0.2	11 000
O/d Forms	Formwork and other concrete materials	5	20	0.2	55 000
O/d Rail.	Railings and guard rails	20	20	0.2	21 000
O/d Deck	Precast deck beams	30	20	0.2	28 500
	Earthworks				
Excav. A	Excavation for abutment A	5	5	0.8	25 000
Excav. B	Excavation for abutment B	5	5	0.8	25 000
Backf. A	Backfill and compact abutment A	4	5	0.8	32 000
Backf. B	Backfill and compact abutment B	4	5	0.8	32 000
	Piling				
Pile A	Drive piles abutment A	8	10	0.8	12 000
Pile B	Drive piles abutment B	8	10	0.8	12 000
Pile 1	Drive piles pier 1	7	10	0.8	12 000
Pile 2	Drive piles pier 2	7	10	0.8	12 000
	Concrete, formwork and reinforcement placing				
Conc. A	Abutment A	10	15	0.5	16 500
Conc. B	Abutment B	10	15	0.5	16 500
Conc. 1	Pier 1	8	15	0.5	16 500
Conc. 2	Pier 2	8	15	0.5	16 500
	Precast concrete placing				
Deck 1	Deck from abutment A to pier 1	4	20	0.85	8000
Deck 2	Deck from pier 1 to pier 2	4	20	0.85	8000
Deck 3	Deck from pier 2 to abutment B	4	20	0.85	8000
	Finishing activities				
Surface	Placing surface and fixing railings and guard rails	10	15	0.8	18 500
Approaches	Approach roads	5	10	0.8	5000
Slopes	Slopes to approach roads	2	10	0.8	13 000
End	Wind up	1	10	0.8	1000

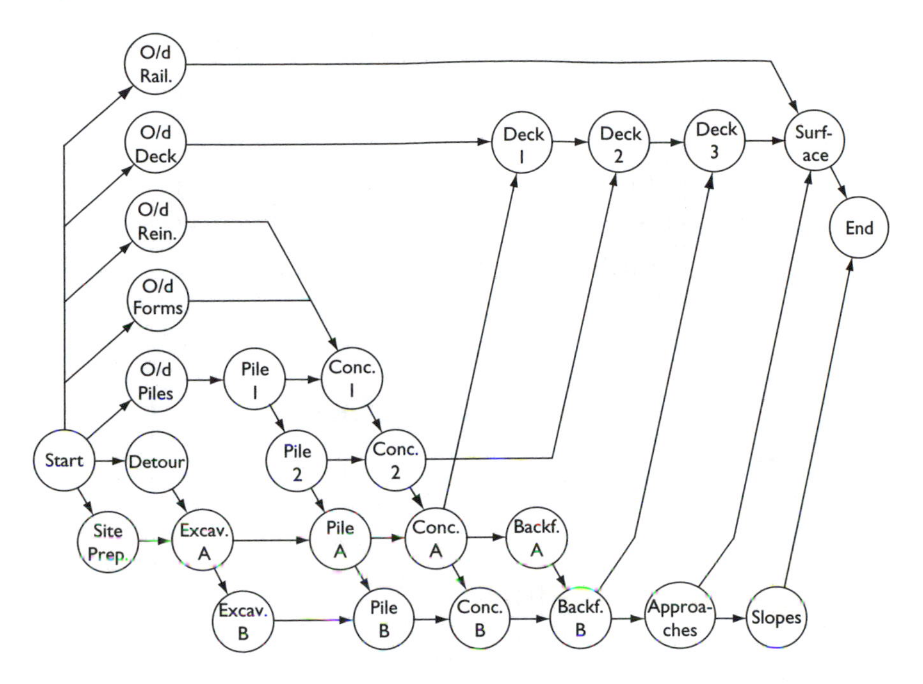

Figure 7.13 Example three-span bridge project.

The simulations are described in Al-Sadek and Carmichael (1992). Simulation output constituted the following:

- A histogram of the duration of the project. This is used to evaluate the probability of completing the project by a certain date. Confidence intervals may also be established from this histogram.
- A histogram of the project cost. This is used to evaluate the probability of completing the project on a certain budget. Confidence intervals may also be established.
- Criticality indices.

For 50 000 simulations, Figure 7.14 shows the histogram of the project duration, and Figure 7.15 shows the histogram of the project cost.

Table 7.7 lists the criticality indices. The first column corresponds to Option (i) where the project latest finish time equals the project earliest finish time. The second and third column of values correspond to Option (ii) with, respectively, a 90-day planned finish and an 80-day planned finish. The fourth column of values corresponds to Option (iii) where the project latest finish time equals the average value of the project duration (111.2 days by simulation).

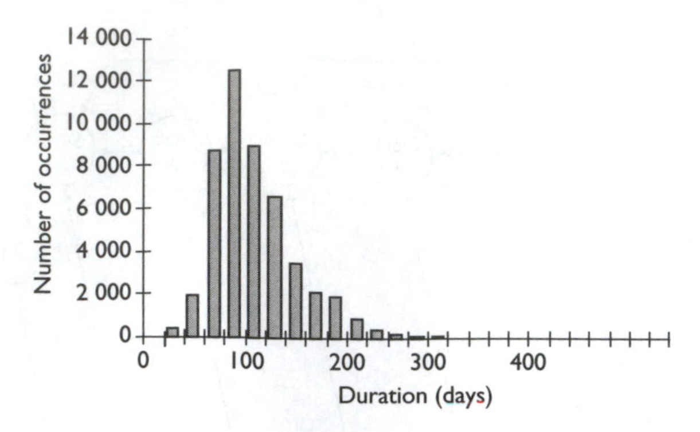

Figure 7.14 Example histogram of the project duration.

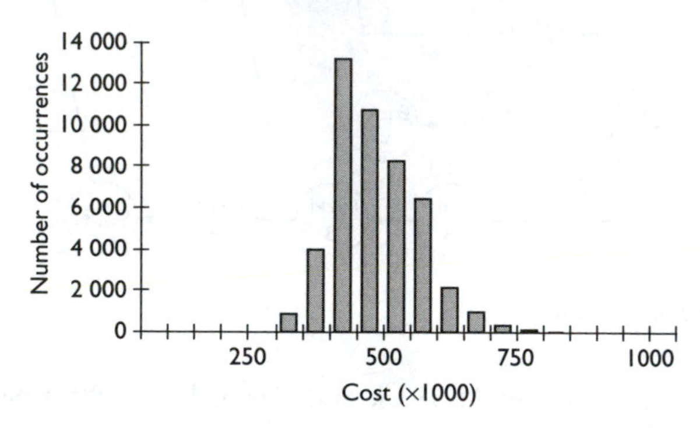

Figure 7.15 Example histogram of project cost.

Variance reduction

The final step in Monte Carlo simulation involves collecting relevant statistics on important items such as project completion times. This is basically a sampling process, and sampling errors decrease as the sample size increases. Monte Carlo simulation also permits the use of *variance reduction techniques* which reduce errors or variances, and these can be used in place of increasing the sample size from increased simulations.

To demonstrate this, consider the project completion time, T. For independent sampling, the variance of the estimate $\tilde{T}$, for d samples, is

$$\begin{aligned}
\mathrm{var}[\tilde{T}] &= \frac{\mathrm{var}(T_1) + \mathrm{var}(T_2) + \cdots + \mathrm{var}(T_d)}{d} \\
&= \frac{d\,\mathrm{var}(T)}{d^2} \\
&= \frac{\mathrm{var}(T)}{d}
\end{aligned}$$

Table 7.7 Criticality indices for example project

Activity	*Criticality indices (%)*			
	Option (i)	*Option (ii)*		*Option (iii)*
		80 day	*90 day*	
Establishment activities				
Start up	100.00	65.04	77.07	41.14
Set up road detour	0.03	11.64	18.91	4.35
Site preparation	2.32	13.34	21.79	4.91
Order and delivery activities				
Piles	83.01	60.54	71.0	38.24
Reinforcement	0.18	14.29	21.87	5.10
Formwork and other concrete materials	0.00	14.53	22.54	5.13
Railings and guard rails	1.47	2.01	3.97	0.69
Precast deck beams	12.99	10.09	15.30	4.17
Earthworks				
Excavation for abutment A	2.35	13.41	21.91	4.92
Excavation for abutment B	0.99	7.46	12.57	2.26
Backfill and compact abutment A	15.49	53.25	62.75	31.08
Backfill and compact abutment B	85.54	60.73	71.60	38.27
Piling				
Drive piles abutment A	30.07	58.53	69.81	36.72
Drive piles abutment B	7.83	54.54	67.68	33.71
Drive piles pier 1	83.01	60.54	71.00	38.24
Drive piles pier 2	28.07	59.15	70.49	37.25
Concrete, formwork and reinforcement placing				
Abutment A	77.71	60.28	70.80	37.64
Abutment B	70.05	58.74	70.82	37.56
Pier 1	54.51	59.27	70.02	36.02
Pier 2	64.36	59.64	70.31	36.63
Precast concrete placing				
Deck from abutment A to pier 1	13.56	54.49	68.25	31.43
Deck from pier 1 to pier 2	13.56	54.59	68.25	31.43
Deck from pier 2 to abutment B	25.47	62.92	75.27	39.62
Finishing activities				
Placing surface and fixing railings and guard rails	98.48	64.99	77.02	41.14
Approach roads	73.11	60.15	71.37	37.70
Slopes to approach roads	1.52	49.53	61.76	29.25
Wind up	100.00	65.04	77.07	41.14

That is, as d increases, the variance decreases and the estimate $\tilde{T}$ gets closer to the true value $E(T)$. Doing calculations for large d may, however, not be feasible and alternative methods may be invoked to improve the efficiency of the computations. Consider one such method, namely that of antithetic variables (for example, Graver and Thompson, 1973).

For every uniformly distributed random number v_i, an additional random number, $1 - v_i$, is generated. Sets of random numbers $v_i, i = 1, 2, \ldots$, give rise to T as before. Sets of random numbers $1 - v_i, i = 1, 2, \ldots$, give rise to T' in a similar fashion. T and T' are antithetic realisations.

The estimates become,

$$\tilde{T}^* = \frac{1}{d}\left(\frac{T_1 + T_1'}{2} + \frac{T_2 + T_2'}{2} + \cdots + \frac{T_d + T_d'}{2}\right)$$

and

$$\text{var}(\tilde{T}^*) = \frac{1}{2d}[\text{var}(T) + \text{cov}(T, T')]$$

The covariance term is negative because T and T' are negatively correlated (antithetic). Large values of T will generally be associated with small values of T' and vice versa, because of the underlying choice of the uniform random numbers; if v_i is large, then $1 - v_i$ is small and vice versa. The negative covariance term should lead to a smaller variance estimate for the project completion time than by direct random sampling.

References

Al-Sadek, O. and Carmichael, D. G. (1992), On Simulation in Planning Networks, *Civil Engineering Systems*, Vol. 9, pp. 59–68.

Ang, A-H. S. and Tang, W. H. (1984), *Probability Concepts in Engineering Planning and Design*, Vol. II, New York: Wiley.

Graver, D. P. and Thompson, G. L. (1973), *Programming and Probability Models in Operations Research*, Monterey: Brooks/Cole.

Otnes, R. K. and Enochson, L. (1978), *Applied Time Series Analysis*, New York: Wiley.

Van Slyke, R. M. (1963), Monte Carlo Methods and the PERT Problem, *Operations Research*, Vol. 11, No. 5, pp. 839–861.

Exercises

1. Consider a triangular distribution (probability density function) for an activity's duration as in Figure E7.1a where, t_1 is an optimistic duration, t_3 is a pessimistic duration and t_2 is the most likely duration. Note, there

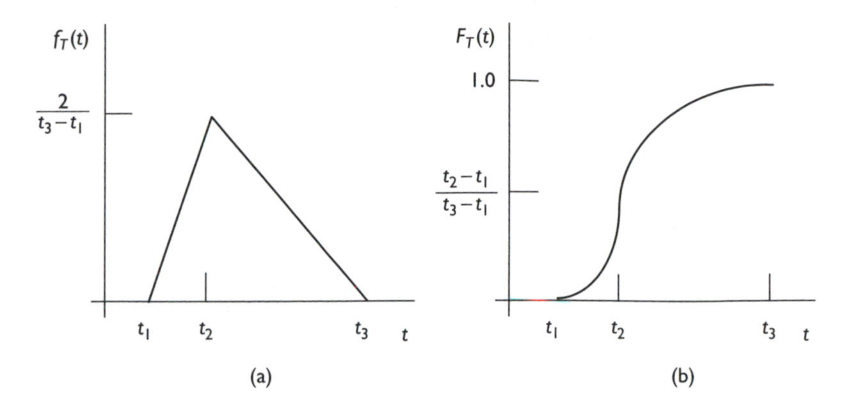

Figure E7.1 PDF and CDF for activity duration.

is a discontinuity at $t = t_2$. Figure E7.1b shows the associated cumulative distribution function.

Show that the relevant transformation between the standard uniform variate V, and T is

$$t = [(t_2 - t_1)(t_3 - t_1)v]^{1/2} + t_1 \quad 0 < v \leq \frac{t_2 - t_1}{t_3 - t_1}$$

$$t = t_3 - [(t_3 - t_2)(t_3 - t_1)(1 - v)]^{1/2} \quad \frac{t_2 - t_1}{t_3 - t_1} < v < 1$$

(*Hint.* You may find the integration of the line equation of the first part of the probability density function for T easier if you first shift the origin of the line to $t = t_1$, and the integration of the line equation for the second part of the probability density function for T easier if you next shift the origin to $t = t_3$.)

2. The following is a FORTRAN listing of a main program and subroutine based on the pseudorandom number generator,

$$r_{i+1} = ar_i + c - mk_i$$

where $r_0 = 673$, $a = 17$, $m = 2^{20} = 1\,048\,576$ and $c = 1$.

```
PROGRAM SIM
...
CALL ATRAND(1,X,IR)
...
CALL ATRAND(0,X,IR)
...
END
```

```
      SUBROUTINE ATRAND(K,X,IR)
      IF(K.EQ.0) GOTO 1
      IR = 673
    1 CONTINUE
      IA = 17
      IM = 1048576
      IC = 1
      IK = INT((IA * IR + IC)/IM)
      IR = IA * IR + IC - IM * IK
      X = FLOAT(IR)/FLOAT(IM)
      RETURN
      END
```

Use FORTRAN or convert to a coding of your choice to generate, say, 1000 random numbers between 0 and 1.

Calculate the sample average,

$$\bar{v} = \sum_{i=1}^{1000} v_i$$

and sample variance,

$$s^2 = \frac{1}{999} \sum_{i=1}^{1000} (v_i - \bar{v})^2$$

of these 1 000 numbers. The sample average should be approximately 0.5. The sample variance should be approximately 1/12.

Draw a histogram of these 1 000 numbers. The choice of interval size on the horizontal axis is up to you. How close is the histogram in shape to a rectangle?

3. The following is a FORTRAN listing of a main program and subroutine based on the pseudorandom number generator given in Otnes and Enochson (1978, p. 414).

```
      PROGRAM SIM
      . . .
      CALL TDRAND(1,X,I)
      . . .
      CALL TDRAND(0,X,I)
      . . .
      END
      SUBROUTINE TDRAND(K,X,I)
      IF(K.EQ.0) GOTO 1
      I = 783637
```

```
 1  CONTINUE
    I = 125 * I
    I = I - (I/2796203) * 2796203
    X = FLOAT(I)/2796202.
    RETURN
    END
```

Do calculations similar to the previous exercise.

Without using any involved statistical method, which pseudorandom number generator, ATRAND or TDRAND, in your opinion gives the better approximation to uniform random numbers?

4. Search the simulation literature for pseudorandom number generators other than the one given above.

Linear projects

Introduction

Many projects are 'linear' in nature and repeat the same sequence of tasks many times throughout the project lifetime. An example is the construction of multistorey buildings where the same trades (for example, involving the plumber, electrician, ceiling fixer and painter in sequence) are repeated floor by floor. Other examples include pipeline and railway construction and estate housing projects.

This linearity is exploited by cumulative production plot-based approaches (activity level representation). The unit of production may be, for example, kilometres, storeys or material usage. Such a plot shows an activity's production up to any time or distance; the slope of the plot represents the rate of production (Figure 7.16).

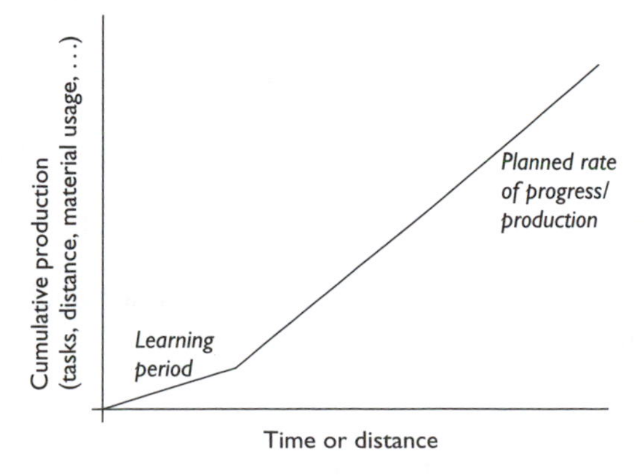

Figure 7.16 Example cumulative production plot.

Some variants on Figure 7.16, at the activity level, are:

- The axis labelling may be interchanged, with the independent variable on the vertical axis
- The diagram may be drawn down the page
- The line plot may be given a finite width where work occupies durations of significance.

Cumulative production plots contain bar chart information integrated. It is the additional *vertical scale* that makes the cumulative production plot attractive, compared to a bar chart. The computational path to the cumulative production plot should be no different to that of a conventional bar chart, namely via a network. A network helps to clarify the logic and dependencies between the activities.

Line of balance thinking applies to linear type projects; this is a process-oriented approach which has its origins in the manufacturing industry.

In cumulative production plots, a *resource schedule* is one determined by resource numbers (quantity), while a *parallel schedule* is one determined by a desired activity production rate. Activity production rate and activity resource usage are directly related; activity production rate can be increased/decreased by the use of more/less resource numbers.

Parallel scheduling involves lines (representing activities) being drawn parallel to each other, with or without a buffer between them. To ensure such activity production rates, the resource numbers are increased (or decreased) accordingly. Such a scheduling practice might be adopted if there is a need for a reduced project duration or a desired completion date. Parallel scheduling will lead to shorter project durations but this is at the expense of using additional resource numbers.

Resource scheduling is where the schedule is determined by the available or usual resource numbers. At the planning stage, the start or finish times of activities are adjusted in order to avoid interference between activities. Such a scheduling practice might be adopted where the least cost solution is sought irrespective of the project duration. Activities with the smaller line slopes can be speeded up by the employment of extra resource numbers and this will have the effect of reducing the overall project duration, but at an additional cost.

Separation or *buffer time periods* are selected to prevent interference between adjacent activities, and are represented by gaps between the activities in a cumulative production plot.

Concurrent activities may be overlapped but this leads to a cumulative production plot presentation that may be harder to read.

Learning

Repetition of work generally leads to later activities being done in shorter durations than the same activities done earlier. Workers become more adept at their jobs with experience. Such behaviour is included in a study of learning and learning curves. Taking learning into account will bend the lines in the cumulative production plot to being closer to vertical as the number of repetitions grows.

Examples

Example

Consider two activities, in a project, namely 'excavate trench' and 'lay pipe'. Figure 7.17 shows the bar chart and associated time-chainage chart, where the activities are proceeding at the same rate.

Should the second activity proceed at a faster rate than the first, the situation in Figure 7.18 might occur.

To avoid the situation that occurs in Figure 7.18, namely that the program says to lay the pipes before a trench has been excavated, a finish-to-finish (F/F) relationship might be used, instead of a start-to-start (S/S) relationship, as shown in Figure 7.19.

Example

Consider an earthmoving project involving cut, and installation of drainage. The bar chart and time-chainage chart may look like Figure 7.20.

Example

The excavation of drains is sometimes carried out from the bottom of a hill upwards, or roads are constructed simultaneously from each end. In

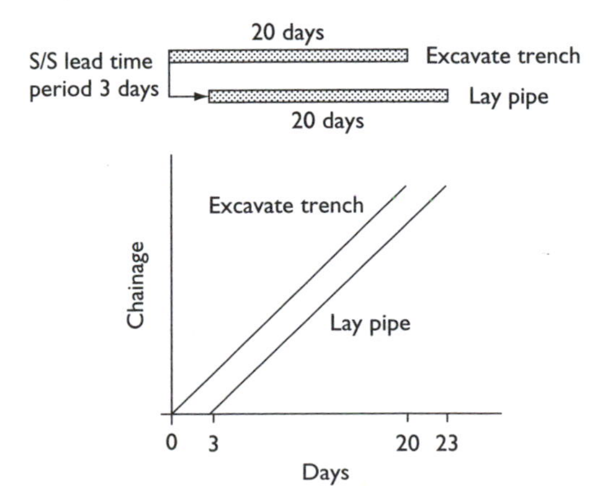

Figure 7.17 Activities advancing at the same rate, and S/S relationship.

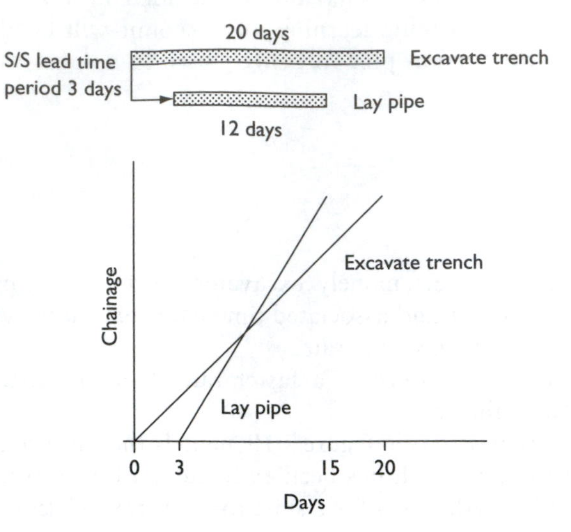

Figure 7.18 Activities advancing at different rates, and S/S relationship.

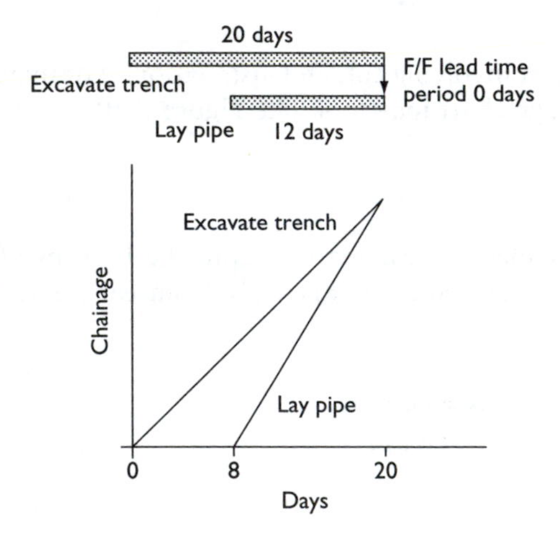

Figure 7.19 Activities advancing at different rates, and F/F relationship.

such cases the time-chainage charts might look like Figure 7.21. Note the orientation of the chart has been changed from the earlier examples, in order to also show the ground terrain.

Example

A complete picture of road construction showing cuts and fills, roadworks, fences etc. may be put on a time-chainage chart, for example Figure 7.22.

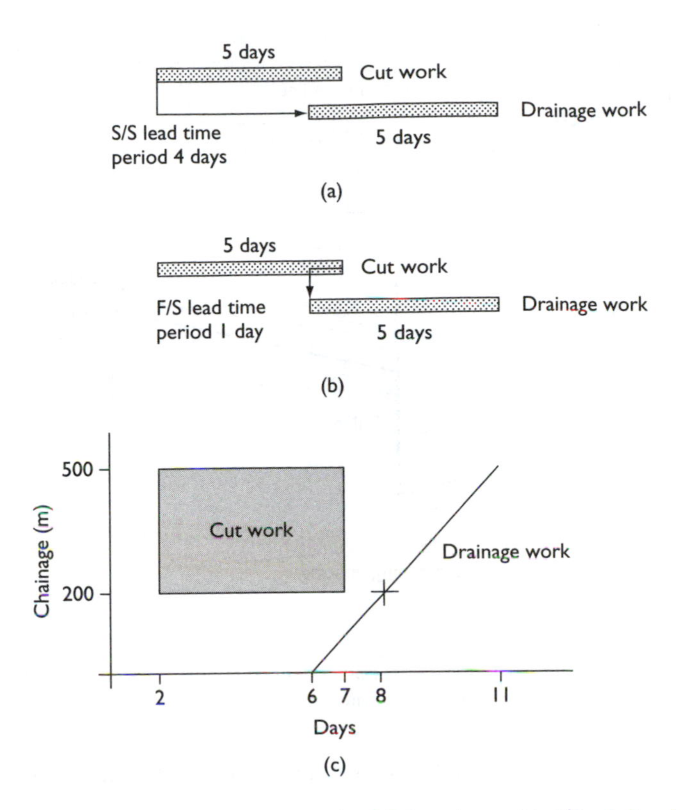

Figure 7.20 Earthmoving example. (a) Bar chart with S/S relationship; (b) Bar chart with F/S relationship; (c) Time-chainage chart.

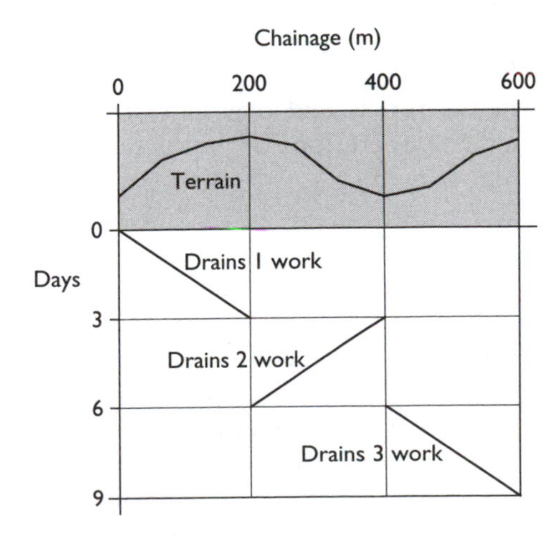

Figure 7.21 Drain construction example.

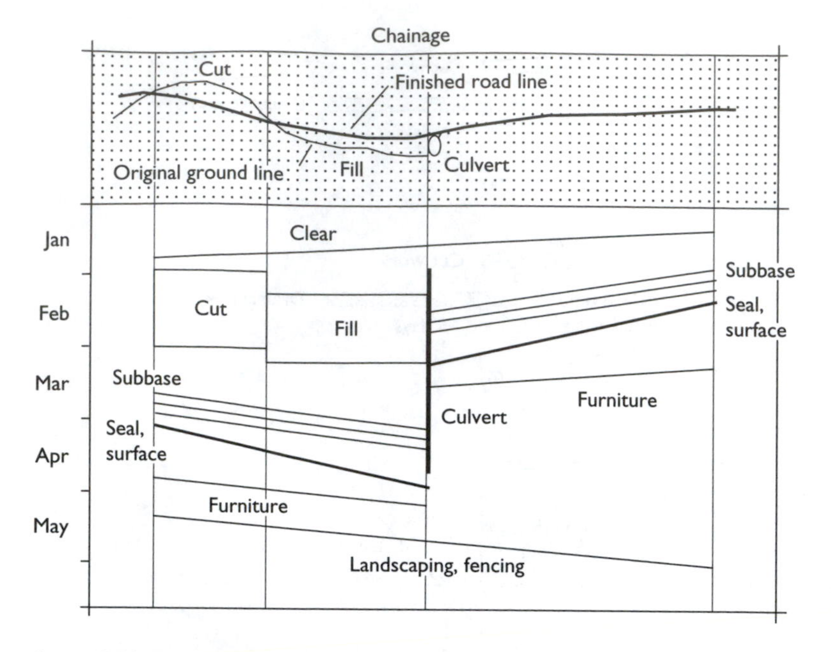

Figure 7.22 Time-chainage chart for road construction. (Activity names have been abbreviated because of space limitations – work items are implied.)

Similar diagrams can be used for railway construction, pipeline construction and so on.

Case example – Aircraft engines

In an organisation involved in the repair and overhaul of aircraft engine turbine blades and vanes, the operations or production work is divided into two main sections, namely:

- Manufacturing Operations
- Special Process Operations.

The Manufacturing Operations section is divided into various product cells, namely the high pressure turbine blades, high pressure turbine vanes, low pressure turbine blades and low pressure turbine vanes. Each product cell is arranged in a U-shaped formation. The principle behind this production concept is to have as many operations as possible located within the cells to cut down on operator and parts movement, which are considered unnecessary and hence labelled as waste. The idea is to have the parts move from one station to another in a 'pull system'. That means output is only what is required by the next customer or next operation.

In earlier days, the production system was similar to a 'push system', whereby product was pushed through the process in batches, and quantities were large enough to meet present and future customer demand and compensate for identified and unidentified problems in the manufacturing process. The push system had no discipline on inventory, but regarded more as always better.

In the 'push system', processes were broken up into specific groups, for example welding, blending, machining and grit-blasting. With the implementation of the 'pull system', all common equipment was separated and placed into each one's respective product cell. At the end of the day, more equipment was required in order to have a complete self-sustaining cell with smooth movement of product from station-to-station. What this meant was that the purchasing of additional equipment did not have to go through any proper justification study. The only reason given was that the cell needed it. Utilisation was never an issue.

The Special Process Operations consisted mainly of batch processes undergoing for example, brazing, coating, plasma spraying and heat treatment. The 'pull system' was avoided. The justification for purchasing is quite different to that of the Manufacturing Operations section. Because most of the equipment in the Special Process Operations section is highly capital intensive, a thorough feasibility study is required. The selection of equipment is based on technical as well as economic considerations. Utilisation is the most important criterion. The company cannot afford to have an expensive 'white elephant' sitting around, if it can be avoided.

Continuous components

Time-chainage charts

Time-chainage charts are seen to provide some information additional to that on connected bar charts. Time-chainage charts contain lines and boxes. Within boxes no other activities may intrude. The charts show:

- The sequence of activities over time
- The location of activities in space
- The direction of production/work
- Activity production rate
- Buffers (intervals) between activities
- Float.

Example

Consider a pipelaying operation which involves the following major activities (Carmichael and Bird, 1992):

Clear and Grade
Excavation
Stringing
Welding
Backfilling
Restoration.

The activity Clear and Grade involves the clearing of the pipe route of obstacles for construction (clears trees and scrub, opens fences, installs temporary gates, strips and windrows top soil for later spreading during restoration) and establishing acceptable construction working grades.

Excavation of the trench follows the Clear and Grade activity. Typically the trench will be, say, 300 mm wider than the diameter of the pipe and provide for cover over the pipe. The bottom of the trench is bedded in sand in preparation for receiving the pipe.

Stringing the pipe follows trenching; the possibility of damage to the insulating coating applied to the pipe is an important consideration.

The important Welding operation follows the stringing. The Welding operation includes the radiographic testing of the welds to prove their integrity prior to joint coating and lowering the pipe into the ground.

When the welds have been cleared and the joints coated, the pipe is lowered into the trench and the trench backfilled. Backfilling sees fine-grained padding material placed around the pipe, and then the return of the excavated material.

Restoration of the pipe route proceeds after backfilling and results in the disturbed areas being returned to as near their original undisturbed contours as possible. Top soil windrowed during the clearing and grading is spread across the right of way and the brush and timber cut during clearing is also spread across the work area to help stabilise and promote rapid regeneration.

Pipeline construction can be likened to a production line, the only difference being that in this case the 'factory' moves. This movement may be several kilometres per day.

Each of the above activities is carried out sequentially and in order that a disruption does not stop the whole production train, separations (buffers) equivalent to a few day's production are maintained between activities.

As each section of the pipeline is backfilled and restored, the pipe is pressured to test both the structural strength and leak tightness. The pipe is then commissioned and handed over to the operating organisation.

Using a network diagram, the activities would be represented as in Figure 7.23a. Should there be concern about the durations to perform these

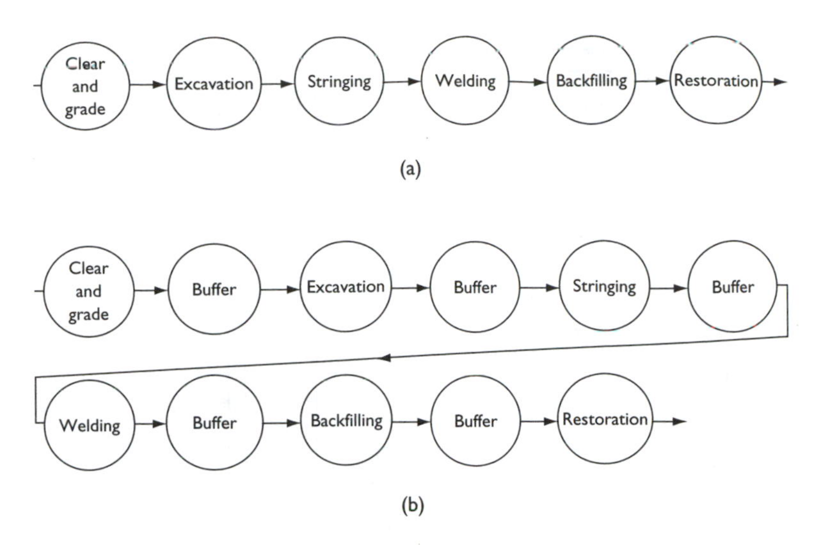

Figure 7.23 Network representation of pipelaying project. (Activity names have been abbreviated because of space limitations – work items are implied.)

activities, then buffers may be inserted into the diagram as in Figure 7.23b. The buffers are contingencies and allow for disruptions. For this example, it is estimated that each of these activities will take 20 weeks and start 1 week after the previous activity. An allowance of 2 weeks is made to separate following activities. Progress is targeted for 3 kilometres per week, with a cycle of 4 weeks work followed by 1 week leave. This information is transferred onto Figure 7.24 which has axes of cumulative production and time.

Figure 7.24 assumes that production for each of the activities progresses at the same constant rate. Should this not be the case, then the lines for each activity will not be parallel (for example, Figure 7.25). The cumulative production plots then assist in working out the required resource numbers for each activity in order that following activities are not interfered with, while at the same time ensuring the necessary target production. This is a common way that cumulative production plots are developed – the estimated production rate for each activity is used to define the slope of that activity's line on the chart. It is the reverse of the procedure as illustrated in Figure 7.25 where the slope of the activity's line establishes what production and hence resource numbers are needed for that activity.

Discrete components

Cumulative production plots take on a slightly different appearance where discrete components are involved, rather than something continuous like a pipeline, railway or road. For a bar chart such as Figure 7.26a, joining

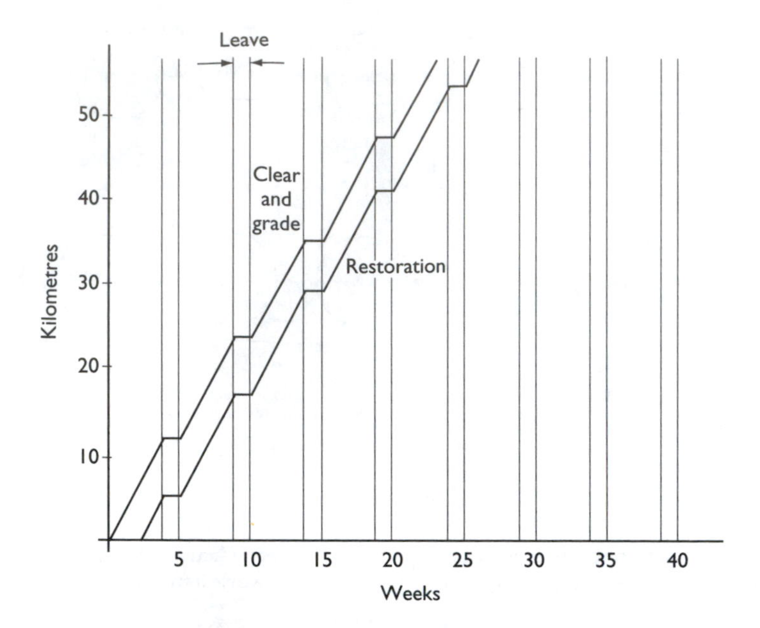

Figure 7.24 Schedule determined by production (parallel schedule). Activities of Excavation, Stringing, Welding and Backfilling not shown for clarity. (Activity names have been abbreviated because of space limitations – work items are implied.)

the bars (that is, 'integrating' the bar chart) and changing the vertical axis gives the cumulative production plot of Figure 7.26b. All the work for each discrete component (in this case a floor of a building) is located on the one line across the page.

When 'integrated', a bar chart becomes the same as a time-based cumulative production plot. Cumulative production plots, bar charts and multiple activity charts contain essentially the same information. It is the additional *vertical scale* that makes the cumulative production plot attractive, compared to a bar chart. The computational path to the cumulative production plot should be no different to that to a conventional bar chart, namely via a network.

Example

An alternative drafting form is illustrated in the following example. The simplified example involves the construction of a bridge represented by three activities – place piers, place beams and place slab – with the sequence repeated tenfold to finish the bridge. Durations for each of these three activities are 9 days, 6 days and 5 days respectively based on using one team of workers for each activity. With buffers of 2.5 days between the

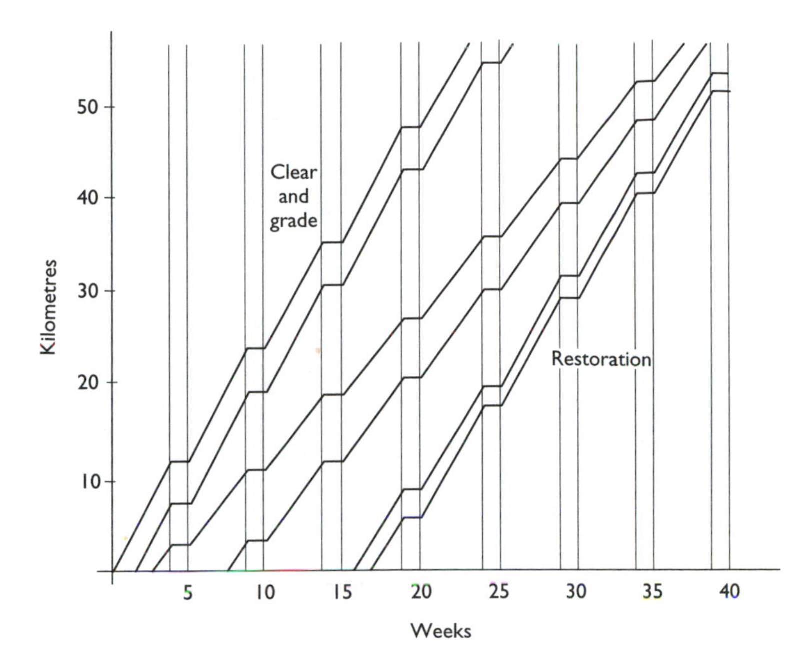

Figure 7.25 Schedule determined by resource numbers (resource schedule). (Activity names have been abbreviated because of space limitations – work items are implied.)

first and second activities and between the second and third activities, this gives a total duration of 25 days for each of the 10 bridge spans.

Assuming that the target completion duration of the total bridge is 47.5 days and each of the three activities is to be carried out at the same production rate, then the cumulative production plot of Figure 7.27a follows. Resource numbers are acquired to achieve the planned activity production rates. From Figure 7.27b four teams of workers are required for the pier work, three teams of workers are required for the beam work and two teams of workers are required for the slab work.

The alternative approach to the scheduling is via the resource schedule (schedule determined by resource numbers) rather than the parallel schedule (schedule determined by production). Here the resource numbers available are first determined and the corresponding activity production rates that can be achieved with these available resource numbers are then calculated.

For example, assume that in the placing of the piers, two teams are employed. The first team would start on the first bridge span on day 11.5 and finish on day 17.5. The second team could then undertake work on the second bridge span. The third span is due to start on day 16.5 but the first team is not available until day 17.5, completing the third span on day

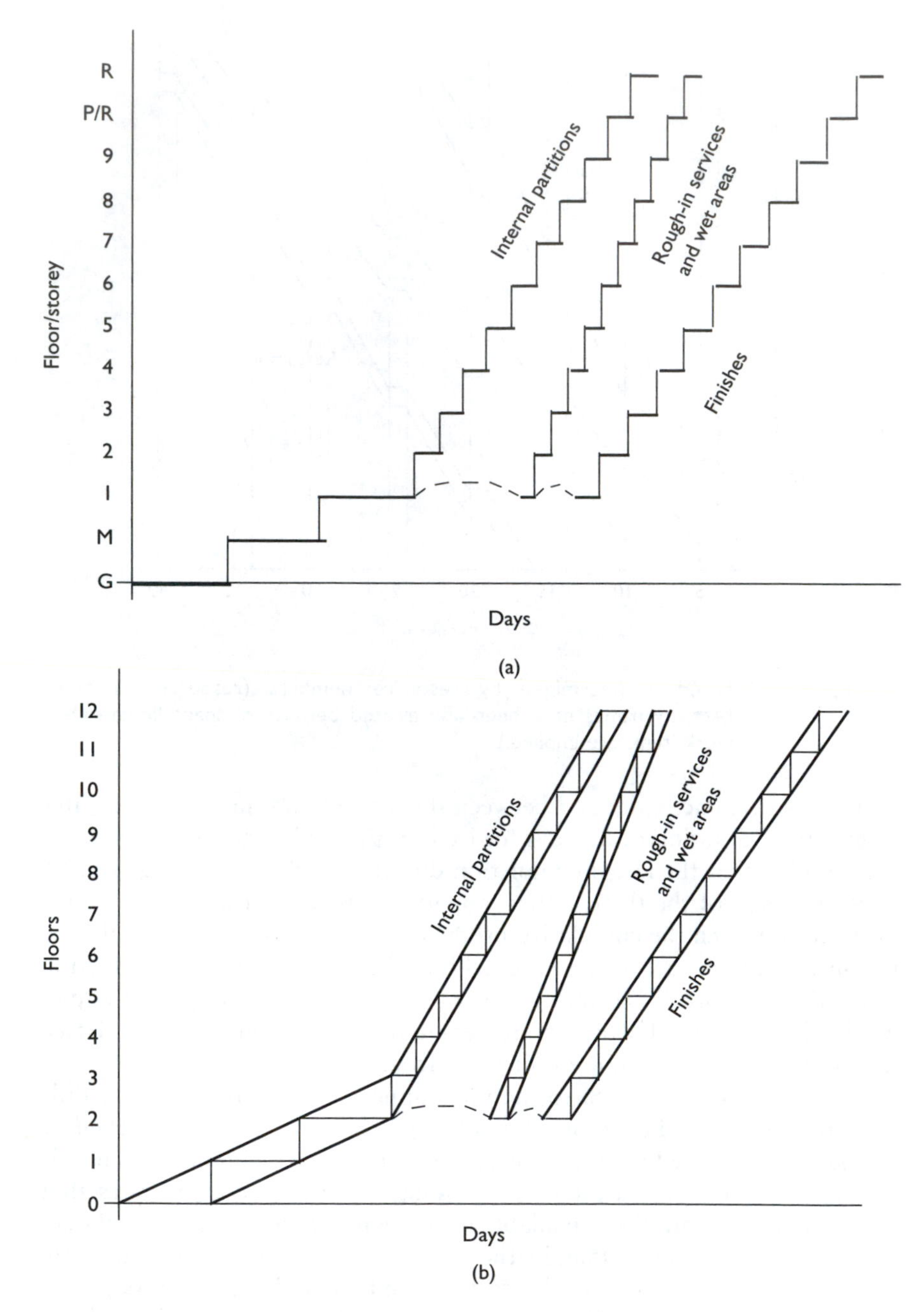

Figure 7.26 Example construction program for internal work for a building. (a) Bar chart; (b) Cumulative version. (Activity names have been abbreviated because of space limitations – work items are implied.)

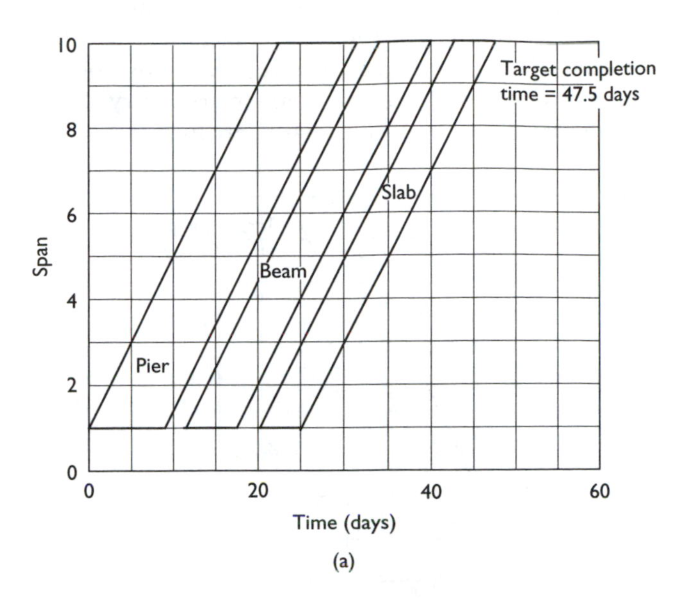

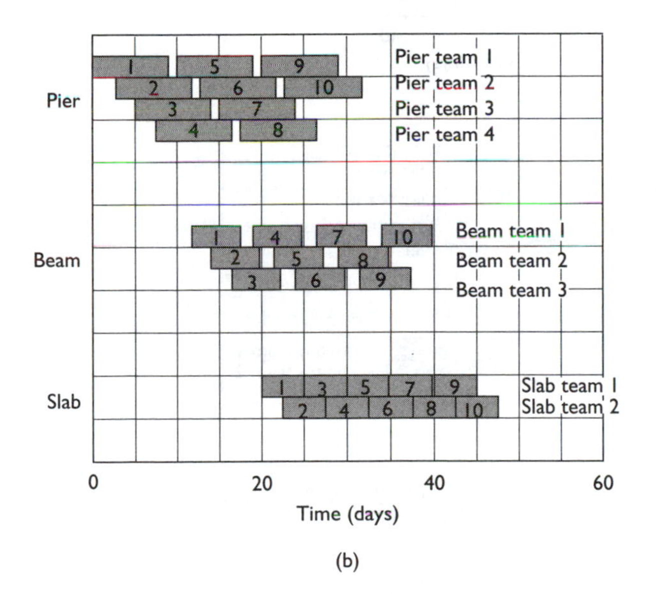

Figure 7.27 Bridge example. (a) Schedule determined by production (parallel schedule). (b) Bar chart for parallel schedule. (Activity names have been abbreviated because of space limitations – work items are implied.)

23.5. The same constraint applies to the second team. That is, the cycle is $23.5 - 17.5 = 6$ days per two spans. Or on average, one span (piers) is completed every 3 days. This is shown in Figure 7.28a. (By comparison Figure 7.27 implies a rate of finishing one span (piers) every 2.5 days.) Figure 7.28b shows the same information in bar chart form.

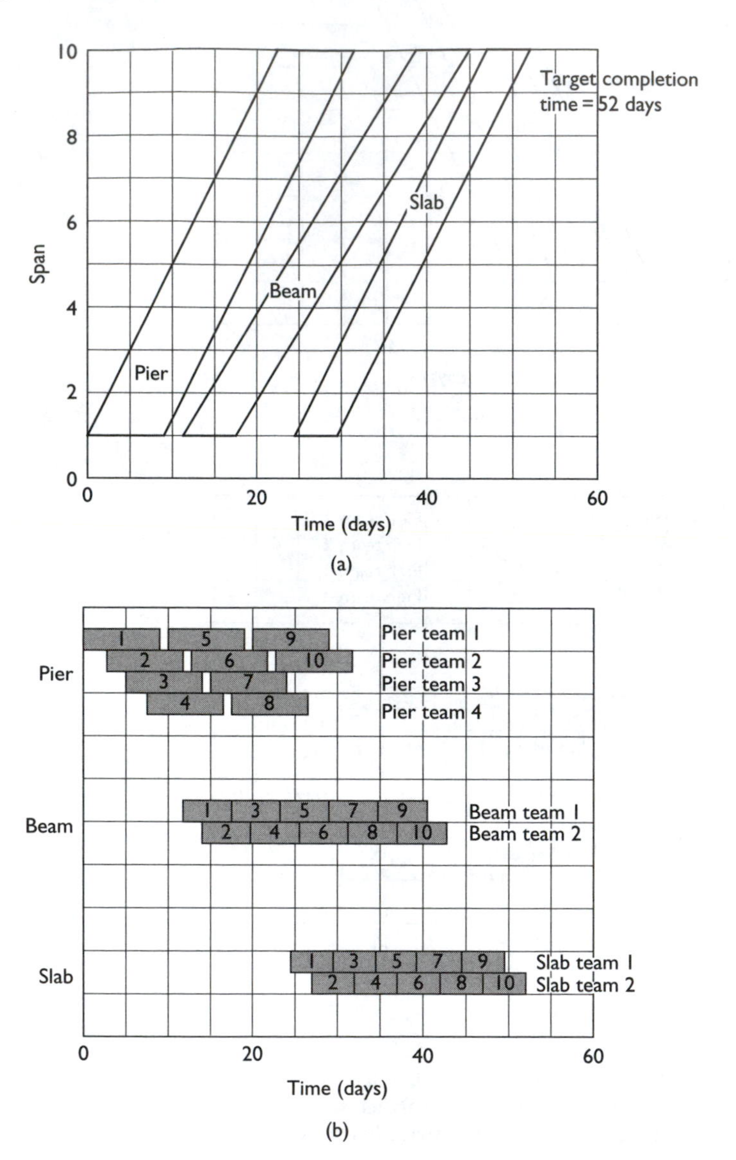

Figure 7.28 Bridge example. (a) Schedule determined by resource numbers (resource schedule). (b) Bar chart for resource schedule. (Activity names have been abbreviated because of space limitations – work items are implied.)

Should three teams be employed to place the piers, the following situation develops. Going backwards in time, the last span of the bridge can be finished on day 40 and started on day 34. That is, the fourth last (or seventh) span can finish on day 34 and start on day 28. (And so on for the seventh last span.) That is, the cycle is $40 - 34 = 34 - 28 = 6$ days per three spans. Or on average, one span (piers) is completed every 2 days. This is shown in Figure 7.29a. Figure 7.29b shows the same information in bar chart form. Interestingly, increasing the team numbers from two to three has no effect on the overall project completion time because the placing of the piers still has to fit in with the other two activities.

Replanning

Replanning can be assisted by overplotting on the cumulative production plots, actual work progress as indicated in Figures 7.30 and 7.31. The rate of production and potential conflict between activities is perhaps not so well seen on bar charts. Controls may be chosen so as to prevent interference between activities.

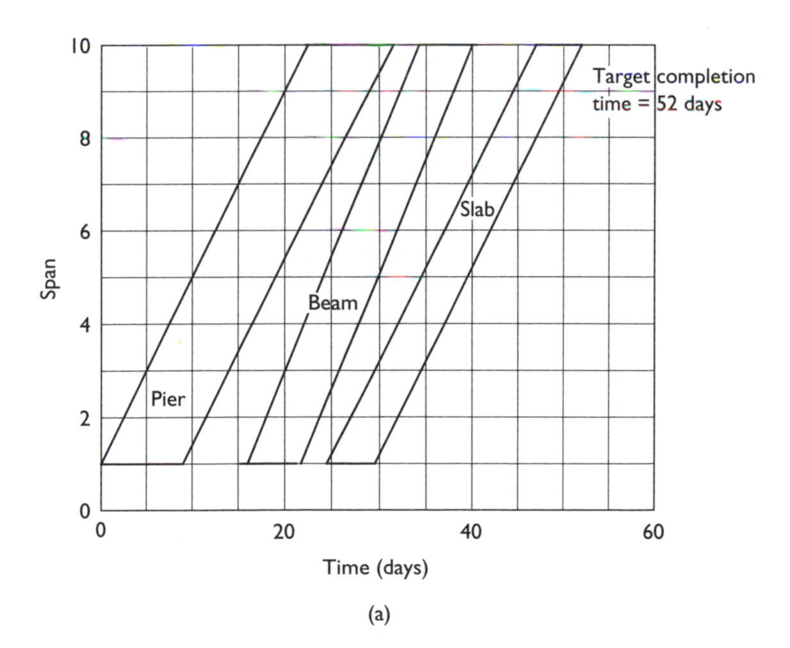

Figure 7.29 Bridge example. (a) Resource schedule; (b) Bar chart for resource schedule. (Activity names have been abbreviated because of space limitations – work items are implied.)

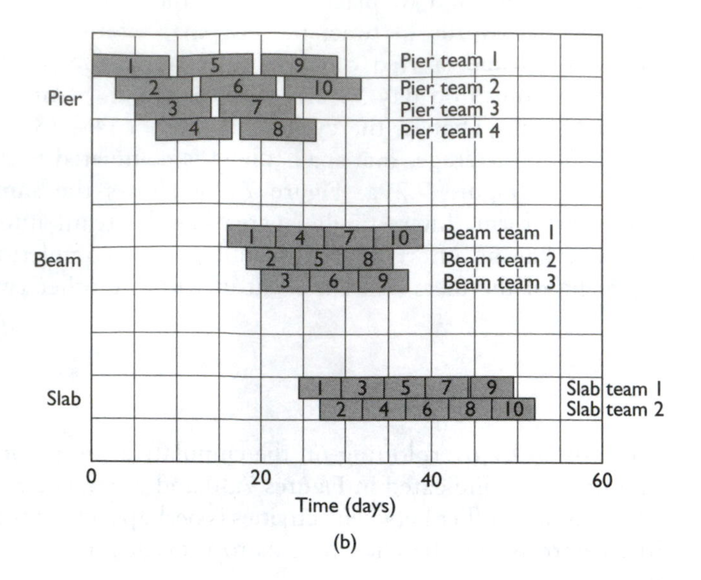

Figure 7.29 (Continued).

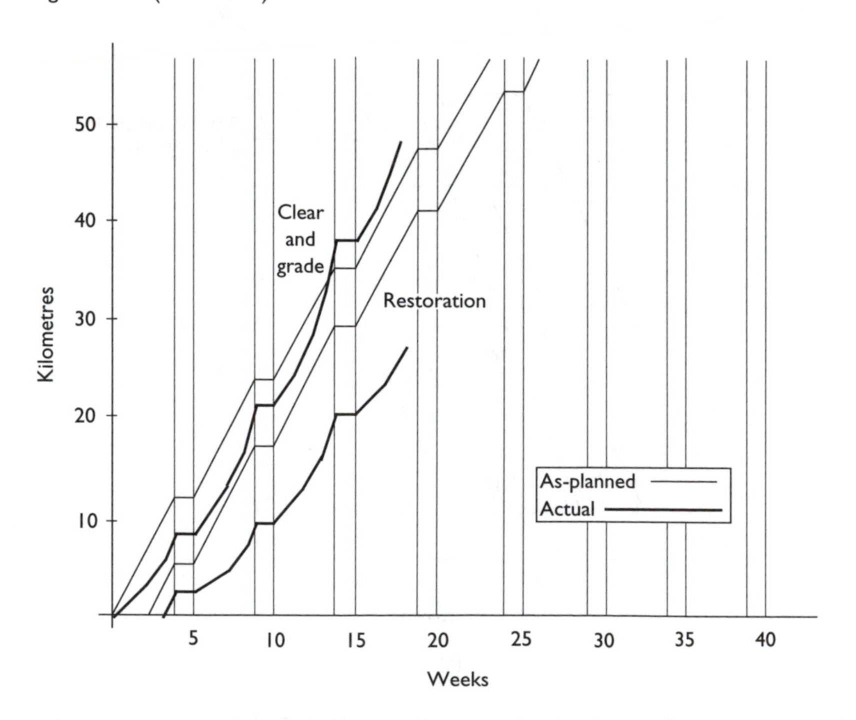

Figure 7.30 Scheduled and actual work; example. (Activity names have been abbreviated because of space limitations – work items are implied.)

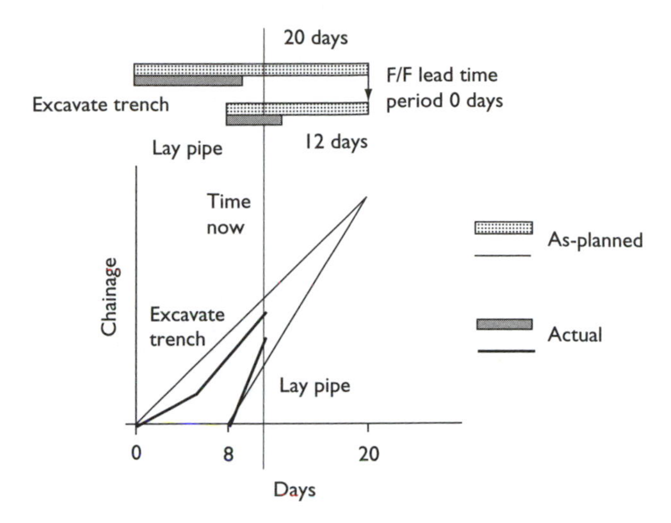

Figure 7.31 As-planned and actual work; example.

Reference

Carmichael, D. G. and Bird, T. R. (1992), *Project Management and Engineering – An Introduction*, Engineering Education Australia (EEA), Sydney.

Exercises

1. Using a similar format to Figure 7.26, what would the program look like where a building was being demolished floor by floor, starting at the top of the building?

2. In repetitive construction, it is important to provide continuity of work for people and plant. For example in multistorey building construction it is important that the different trades when finished working on one storey have work to start on the following storey.
   How does the cumulative production plot ensure that continuity of work is obtained?

3. How do you alter a cumulative production plot in order to reduce non-productive periods?

4. For an activity represented in a cumulative production plot by a single line at 60° to the vertical, how does the shape of the line change if, whenever the work is repeated, a 5% improvement occurs in the duration taken to undertake that work? Sketch graphically what is happening to the cumulative production plot.

Part II

Synthesis treatment

Chapter 8

Multistage planning

Introduction

Planning over the total project duration may be simplified by considering the project to be composed of stages (or phases). The planning problem then becomes a multistage optimal control problem or multistage decision problem. Various optimal control tools are available to assist, including Pontryagin's maximum principle, dynamic programming and mathematical programming.

Controls are chosen for each stage to give a globally optimum solution over the whole project. The controls considered in this chapter relate to resources and resource production rates only, that is the lower level problems in the context of Chapter 5. The work method is assumed given or fixed. The synthesis problem involving work method as controls is very difficult; in the comparative area of structural optimisation, only Michell structure results have evolved.

States are chosen as cumulative resource (or money) usage and cumulative production.

The single objective case is given.

Stages may be chosen to match the type of work, convenient time period divisions in the project, or in an arbitrary sense. Hold or approval points might be used to distinguish stages. Breaking down the project into stages is equivalent to discretising the project duration interval.

This chapter is dealing with planning over the whole project duration using stages. Replanning is also carried out according to stages or at discrete points in time, but this follows from the nature of what replanning is.

Parts of this chapter follow Carmichael (2004).

Stage modelling

Introduction

A (planned) baseline is obtained by solving the deterministic optimal control problem from project start to project end (or the equivalent iterative analysis version). Future uncertain events are converted into equivalent deterministic influences. (There is no need to talk of risk management.) This optimal control problem may be expressed in continuous time or discrete time. The latter is treated here.

Planning over the total project duration

A project may be decomposed into stages. For sequential stages (that is, the staging is based on a time period subdivision; each stage represents a period of time) the decomposition will look like Figure 8.1. Each stage is a subproject.

In Figure 8.1,

N	number of stages
k	stage counter, $k = 0, 1, 2, \ldots, N-1$
$x(k)$	state leading into stage k
$x(0)$	initial state
$x(N)$	final state
$u(k)$	control at stage k

State and control are vector functions, $x = (x_1, x_2, \ldots, x_n)^T$; $u = (u_1, u_2, \ldots, u_r)^T$

Associated with each stage is an objective function $J(k), k = 0, 1, 2, \ldots, N-1$. (See the earlier definition of the term 'objective' used in this book. It is not used in a loose lay person, dictionary or management text sense.)

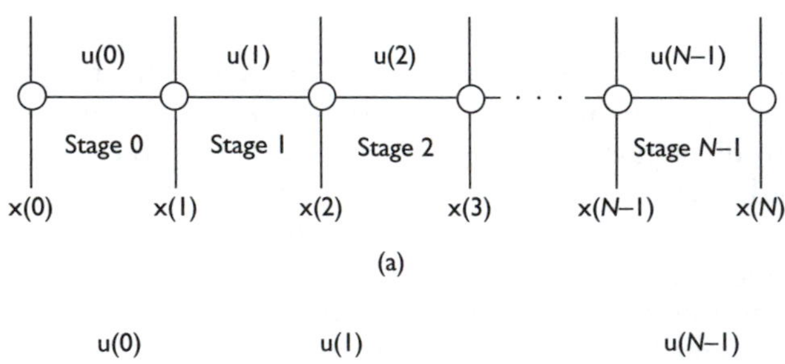

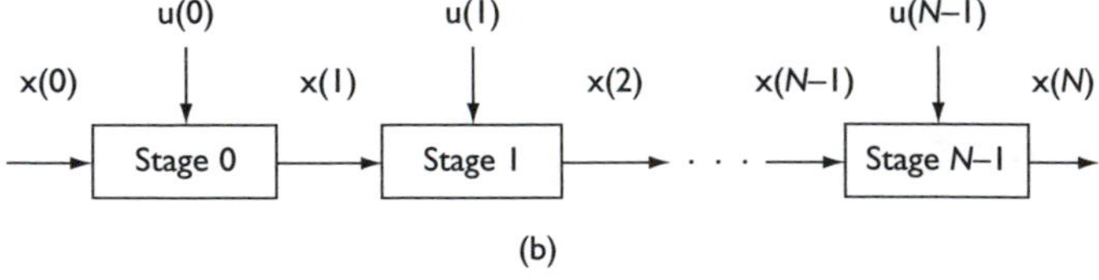

Figure 8.1 Staged (phased) decomposition of a project. (a) 'Activity-on-link' form; (b) 'Activity-on-node' form.

Note that the stage output and the state are the same in this stripped-back project formulation.

Such a formulation is amenable to the application of discrete optimal control systems theory techniques such as dynamic programming. There, an optimisation problem is broken down into subproblems associated with each stage. The optimisation problem is solved by successively going through each stage, from start to finish (or finish to start). Associated with each stage is an objective function which evaluates the worth of each stage control selected from a range of possible alternative controls. The state represents the connection between successive stages; it carries forward information on the system behaviour, such that the state $x(k)$ incorporates information from stages 0 to k–1.

What constitutes the state and control is perhaps the hardest thing to establish when setting up such an optimal control formulation. The state differs from application to application.

> The state at any stage is the minimal amount of information needed to completely determine the behaviour (state) of the system for all other following stages, for any given control. To extend the concept of state to stochastic systems, the state at any stage is regarded as the information that uniquely determines the probability distributions of behaviour (state) at all other following stages. That is, a discrete equivalent of a Markov process.
>
> (Carmichael, 1981)

> The state is information needed from all preceding stages in order to select a control in the current stage without reference to previous controls selected. It represents status information on the project at any point in time.
>
> (Carmichael, 2004)

In line with Chapter 2, the following are selected as the states and controls. (The work method is assumed fixed.)

state(k): cumulative resource usage; the resource usage of stages up to and including stage k–1

cumulative production; the production of stages up to and including stage k–1

control(k): resource numbers chosen at stage k (for each resource type)

production chosen at stage k

In terms of the resource component of the above state and control definitions, cumulative resource usage is in fact the representation of S curves

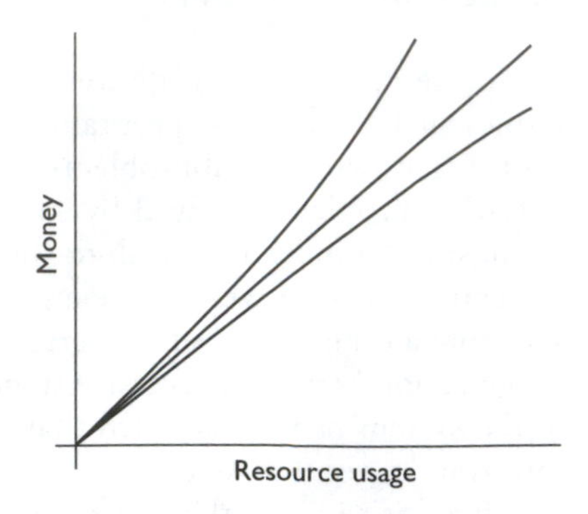

Figure 8.2 Some possible relationships between resource usage and money.

(resources), while the number of states and controls will reflect the number of resource types.

This makes for a very high dimensional problem. The dimension can be reduced if resources are expressed in a common unit of money and aggregated (Figure 8.2).

state(k):	cumulative cost; the expenditure of stages up to and including stage k–1
	cumulative production; the production of stages up to and including stage k–1
control(k):	expenditure chosen at stage k
	production chosen at stage k

In terms of cost, cumulative expenditure is that plotted in the familiar S curve (cost).

$x(k)$ and $u(k)$ are vectors with the above listed components.

State equations

Given the above definitions of state and control, the system equations (model) in state equation (state transformation) form become

$$x(k+1) = x(k) + u(k)$$

where

$x(k)$ state for stage k
$u(k)$ control in stage k

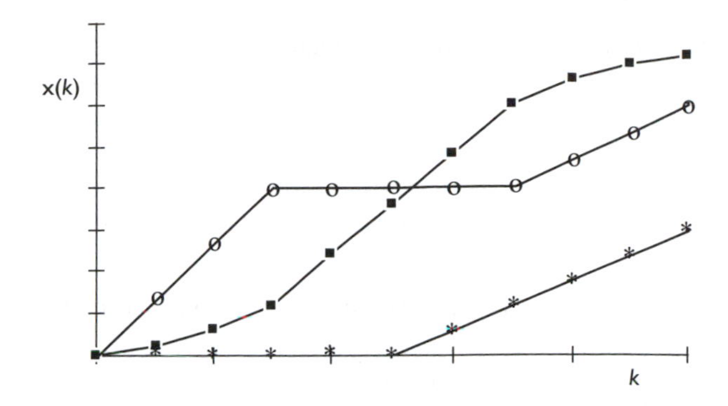

Figure 8.3 Example state trajectories.

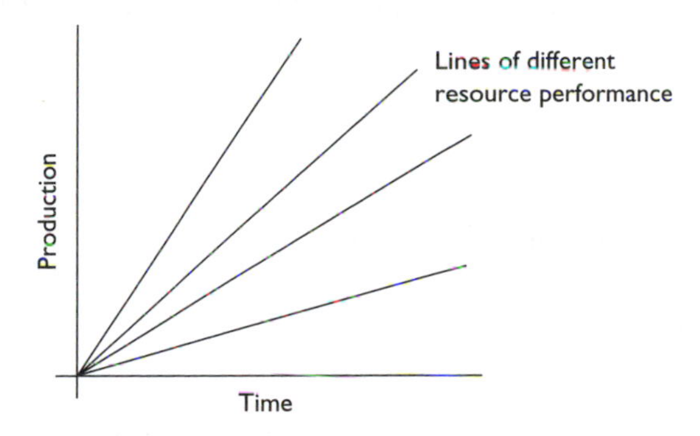

Figure 8.4 Example 'constitutive relationships' for a resource. The slope of a line is a (resource) production rate.

The initial state $x(0) = 0$, but the final state $x(N)$ may not be fixed.

The above state equations are characteristic of discrete (data) control systems.

State trajectories may look something like Figure 8.3.

To track project schedule, the constituent level behaviour of Figure 8.4 is used.

Linearity

Because of the linear state equations, the optimal solution may turn out to be *singular*, possibly even *bang-bang* in nature (Carmichael, 1981). This will depend on the nature of the objective and constraints chosen.

Objectives and constraints

Some example objectives and constraints are given below.

Objectives

There are a number of possible choices of objectives available to the planner. The most obvious would be minimum resource usage/expenditure and minimum duration.

Minimum resource usage/expenditure. For each stage, $J(k) = u(k)$ and

$$\min J = \min \sum_{k=0}^{N-1} u(k)$$

where only the components of u relating to resources are involved.

Minimum duration. For each stage, $J(k) = T(k)$, the stage duration. And

$$\min J = \min \sum_{k=0}^{N-1} T(k)$$

Assuming the constituent level behaviour of Figure 8.4 holds, minimum duration will correspond to maximum production.

$$\min J = \max \sum_{k=0}^{N-1} u(k) = \min - \sum_{k=0}^{N-1} u(k)$$

where only the component of **u** relating to production is involved.

Constraints

Typically, there may be constraints on the overall expenditure and project duration,

$$x(N) \leq \text{some limit}$$

where only the component of x relating to cost is involved, and

$$\sum_{k=0}^{N-1} T(k) \leq \text{some limit}$$

or

$$\sum_{k=0}^{N-1} u(k) \geq \text{some limit}$$

where only the component of u relating to production is involved.

There may also be restrictions on resource usage/expenditure throughout the project,

$$\mathrm{u}(k) \leq \text{some limit} \quad k = 0, 1, 2, \ldots, N\text{–}1$$

where only the components of u relating to resources or expenditure are involved.

Optimisation problem

A generalisation

The above development may be shown to fit within a general optimal control problem statement (Carmichael, 1981). This section is after Carmichael (2004).

System model

The general form of the system (state) equations given in the previous section is

$$\mathrm{x}(k+1) = \mathrm{F}[\mathrm{x}(k), \mathrm{u}(k), k] \quad k = 0, 1, \ldots, N\text{–}1$$

where

$\mathrm{x}(k) = [\mathrm{x}_1, \ldots, \mathrm{x}_n]^T$ is an n-valued vector of the state at stage k
$\mathrm{u}(k) = [\mathrm{u}_1, \ldots, \mathrm{u}_r]^T$ is an r-valued vector of controls at stage k
$\mathrm{F} = [\mathrm{F}_1, \ldots, \mathrm{F}_n]^T$ is a nonlinear vector function.

For given u, and initial and final (terminal) conditions on x, x is completely defined by the state equations.

The equations represent a sequence of transitions from the kth to the $(k+1)$th state, $k = 0, 1, \ldots, N\text{–}1$. The state is assumed to have Markov properties; that is, the state at $k+1$ depends only on the immediately previous state $\mathrm{x}(k)$ and control $\mathrm{u}(k)$.

Remaining optimal control problem components

Objective

A general objective, incorporating the earlier examples, can be written as,

$$\min \mathrm{J} = \mathrm{g}[\mathrm{x}(k)]\Big|_{k=0}^{k=N} + \sum_{k=0}^{N-1} \mathrm{G}[\mathrm{x}(k), \mathrm{u}(k), k]$$

where

G, g are scalar single-valued functions of their respective arguments.

Constraints

Constraints influence the solution by isolating admissible solutions from all possible solutions. Constraints may be given in the form of equalities or inequalities. Constraints may exist on both state and control variables.

Common project constraints include limits on expenditure and limits on resource usage.

The problem

The problem is to determine an admissible sequence of controls $\hat{u}(k)$, $k = 0, 1, \ldots, N-1$ satisfying the system (state) equations, initial and final (terminal) conditions on the state, and any constraints, and minimising J. $\hat{u}(k)$, $k = 0, 1, \ldots, N-1$ is termed the optimal sequence or policy.

The linear–quadratic case

The case involving linear system equations and quadratic objective has a particularly neat solution (Carmichael, 1981).

Singular control

Where the system (state) equations and objective are linear, the solution may be singular or bang-bang in nature.

Determinism

The above is a deterministic formulation. A related stochastic formulation can be given.

Examples

Introduction

Two examples are given below. The first chooses project staging such that the Markov assumptions in the state hold, and this leads to the canonical system equations.

$$x(k+1) = F[x(k), u(k), k] \quad k = 0, 1, \ldots, N-1$$

Relevant solution techniques are Pontryagin's maximum principle, Bellman's dynamic programming and mathematical programming (Carmichael, 1981).

With a linear objective and constraints on the controls, bang-bang and singular solutions may emerge.

The second example has arbitrarily chosen staging. System equations can still be derived but do not follow the canonical form. Here mathematical programming would be the preferred optimisation technique.

Canonical form example

Consider the example network shown in Figure 8.5(a,b). The work method leading to the network is assumed fixed. That is, the project level control is fixed or given.

The project may be subdivided into subprojects or stages. The subdivision which preserves the Markov assumption on the state is given in Figure 8.6(a,b). The lines through the networks in Figure 8.6(a,b) indicate the divisions between subprojects or stages.

Controls $u(i, j)$ are shown adjacent to each activity (i, j) in the activity-on-link diagram, and in the activity boxes in the activity-on-node diagram. Using the notation xj as the state following activity (i, j), the state in and state out (cumulative resource usage/expenditure, cumulative production) for each stage are as follows:

$$x1 \text{ given}$$

$$x4 = x1 + u(1,2) + u(2,4) + u(1,3) + u(3,4)$$

$$x9 = x4 + u(4,5) + u(5,7) + u(4,6) + u(6,7) + u(7,8) + u(8,9) + u(6,9)$$

$$x12 = x9 + u(9,11) + u(9,10) + u(10,11) + u(11,12)$$

These equations belong to the standard (canonical) form

$$x(k+1) = F[x(k), u(k), k] \quad k = 0, 1, \ldots, N-1$$

where

$$N = 3$$
$$x(0) = x1$$
$$x(1) = x4$$
$$x(2) = x9$$
$$x(3) = x(N) = x12$$
$$u(0) = u(1,2) + u(2,4) + u(1,3) + u(3,4)$$
$$u(1) = u(4,5) + u(5,7) + u(4,6) + u(6,7) + u(7,8) + u(8,9) + u(6,9)$$
$$u(2) = u(9,11) + u(9,10) + u(10,11) + u(11,12)$$

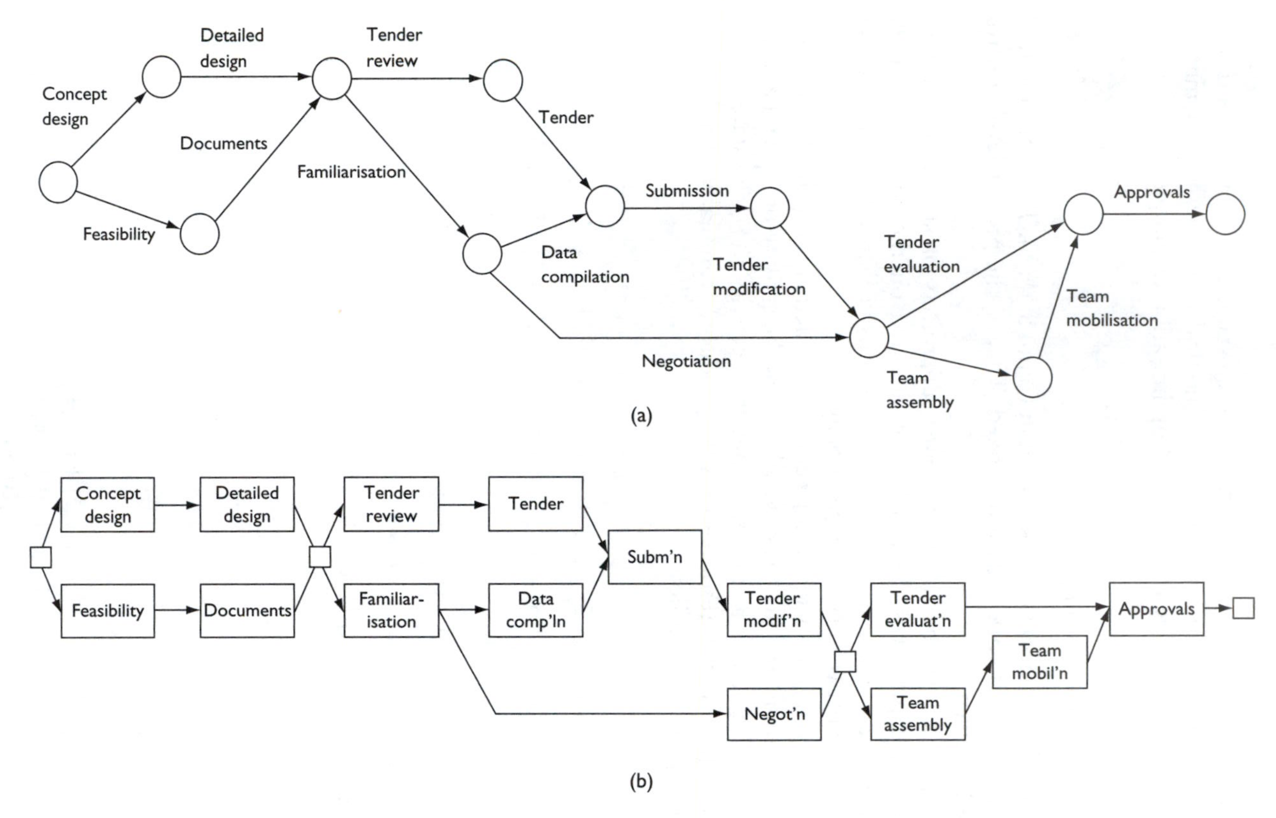

Figure 8.5 Example networks. (a) Activity-on-link diagram; (b) Activity-on-node diagram. (Activity names have been abbreviated because of space limitations – work items are implied.)

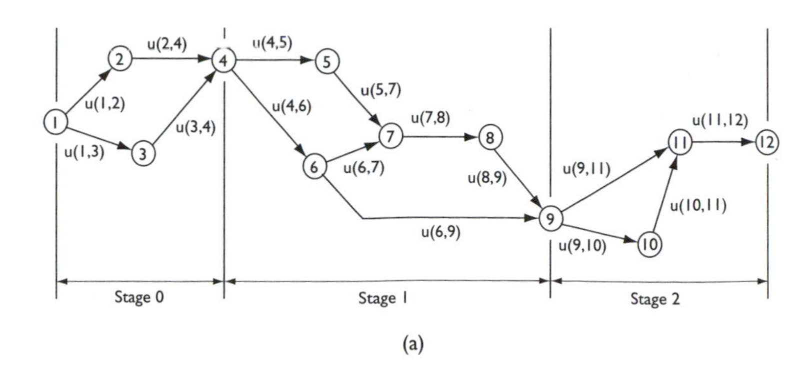

(a)

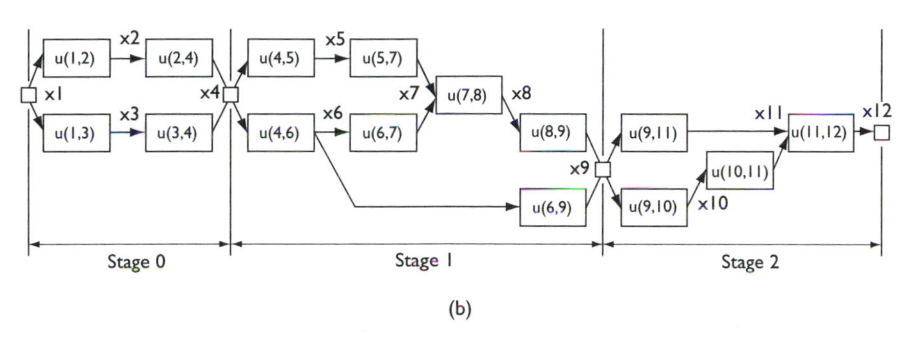

(b)

Figure 8.6 Example staging of project. (a) Activity-on-link diagram; (b) Activity-on-node diagram.

Non-canonical form example

Consider the example network shown in Figure 8.7(a,b). The work method leading to the network is assumed fixed. That is, the project level control is fixed or given.

The project may be subdivided into subprojects or stages. One subdivision has been arbitrarily chosen and is given in Figure 8.8(a,b), but the following development applies to any other similar subdivision. The lines through the networks in Figure 8.8(a,b) indicate the divisions between subprojects or stages; these lines do not have to be vertical. In the activity-on-node diagram of Figure 8.8b, the dummy activities of the activity-on-link diagram, Figure 8.8a, have been added to show the comparable development.

Controls $u(i, j)$ are shown adjacent to each activity (i, j) in the activity-on-link diagram, and in the activity boxes in the activity-on-node diagram. Using the notation xj as the state following activity (i, j), the state in and state out (cumulative resource usage/expenditure, cumulative production) for each stage are as follows:

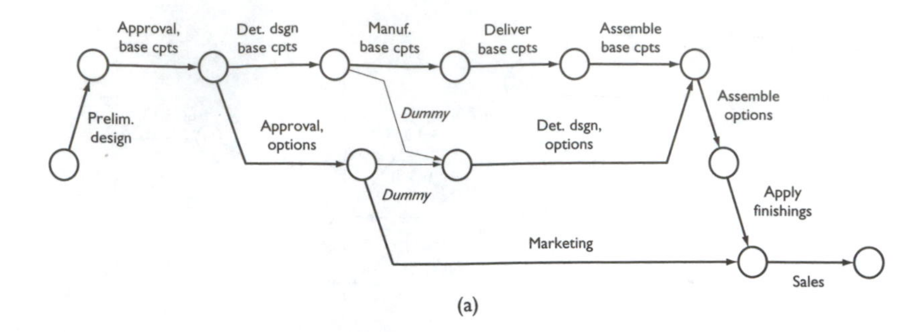

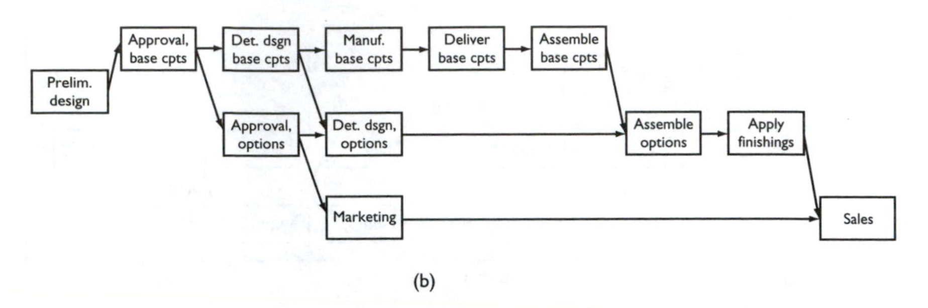

Figure 8.7 Example networks. (a) Activity-on-link diagram; (b) Activity-on-node diagram. (Activity names have been abbreviated because of space limitations – work items are implied.)

Stage 0

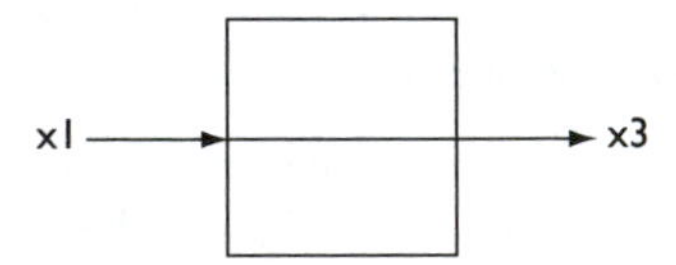

$$x3 = x1 + u(1, 2) + u(2, 3)$$

Stage 1

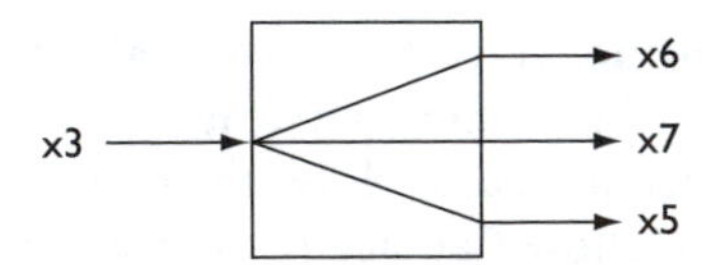

$$x6 = x3 + u(3,4) + u(4,6)$$

$$x7 = x3 + u(3,4) + u(4,7) + x3 + u(3,5) + u(5,7)$$

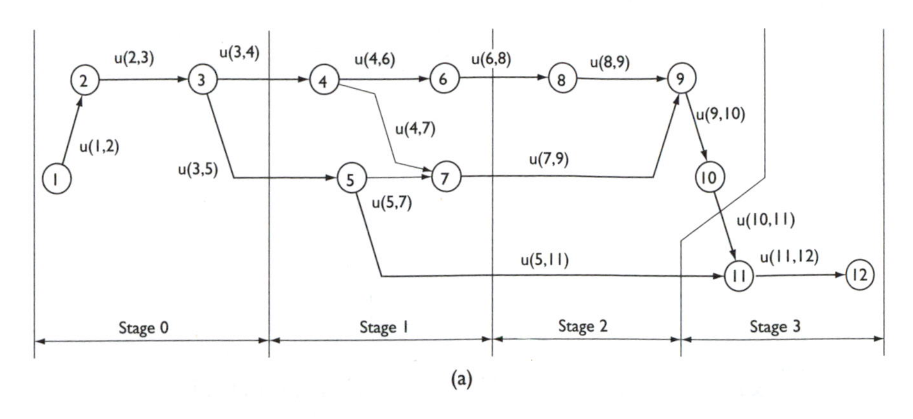

(a)

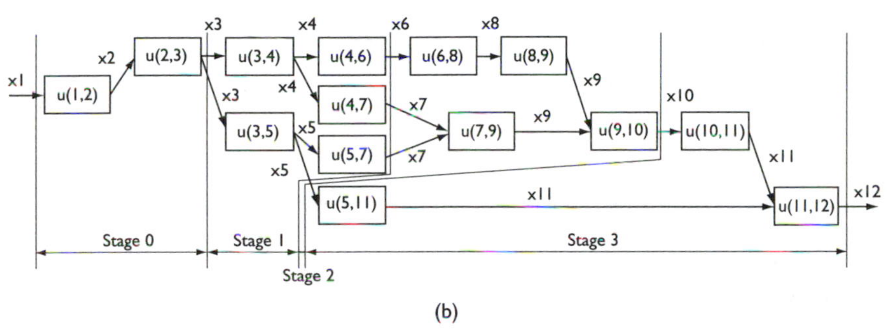

(b)

Figure 8.8 Example staging of project. (a) Activity-on-link diagram; (b) Activity-on-node diagram.

Accounting for redundancies

$$x7 = x3 + u(3,4) + u(4,7) + x3 - x3 + u(3,5) + u(5,7)$$

$$x5 = x3 + u(3,5)$$

Stage 2

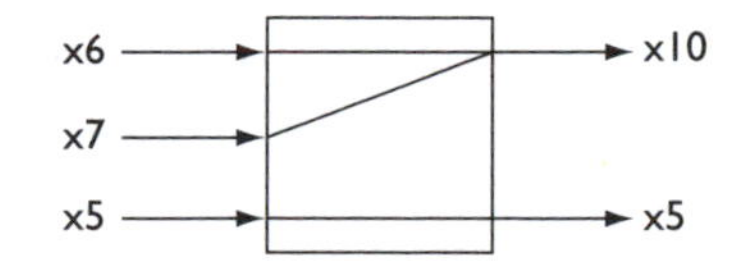

$$x5 = x5$$

$$x10 = x6 + u(6, 8) + u(8, 9) + u(9, 10) + x7 + u(7, 9) + u(9, 10)$$

Accounting for redundancies

$$x10 = x6 + u(6,8) + u(8,9) + u(9,10) + x7 - x3 - u(3,4) + u(7,9) + u(9,10) - u(9,10)$$

Stage 3

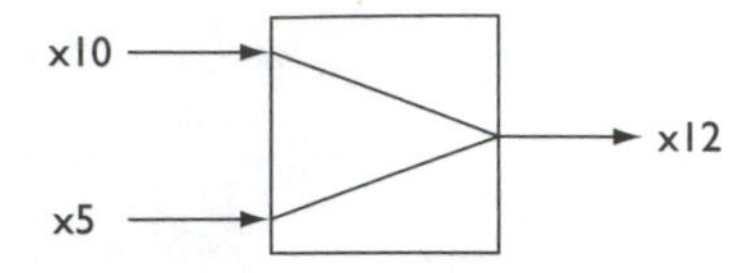

$$x12 = x10 + u(10,11) + u(11,12) + x5 + u(5,11) + u(11,12)$$

Accounting for redundancies,

$$x12 = x10 + u(10,11) + u(11,12) + x5 - x5 + u(5,11) + u(11,12) - u(11,12)$$

These system equations may be collectively written as,

$$\begin{bmatrix} x3 \\ x5 \\ x6 \\ x7 \\ x10 \\ x12 \end{bmatrix} = \begin{bmatrix} x1 \\ x3 \\ x3 \\ x3 \\ x6 + x7 - x3 \\ x10 \end{bmatrix} + \begin{bmatrix} u(1,2) + u(2,3) \\ u(3,5) \\ u(3,4) + u(4,6) \\ u(3,4) + u(3,5) + u(4,7) + u(5,7) \\ u(6,8) + u(8,9) + u(9,10) + u(7,9) - u(3,4) \\ u(10,11) + u(5,11) + u(11,12) \end{bmatrix}$$

with initial conditions x1 given.

The system equations are linear. Controls can be grouped, for example, as they appear in the above system equations, as a tidying up exercise to give,

$$\begin{bmatrix} x3 \\ x5 \\ x6 \\ x7 \\ x10 \\ x12 \end{bmatrix} = \begin{bmatrix} x1 \\ x3 \\ x3 \\ x3 \\ x6 + x7 - x3 \\ x10 \end{bmatrix} + \begin{bmatrix} u1 \\ u2 \\ u3 \\ u4 \\ u5 \\ u6 \end{bmatrix}$$

where,

$$u1 = u(1,2) + u(2,3)$$
$$u2 = u(3,5)$$
$$u3 = u(3,4) + u(4,6)$$
$$u4 = u(3,4) + u(3,5) + u(4,7) + u(5,7)$$
$$u5 = u(6,8) + u(8,9) + u(9,10) + u(7,9) - u(3,4)$$
$$u6 = u(10,11) + u(5,11) + u(11,12)$$

But other options are available for grouping.

The dimension of these system equations can be reduced by reducing the number of stages, and carefully selecting the places where stage divisions occur.

The solution to the associated optimisation (planning) problem might best be performed using mathematical programming.

Chapter 9

Synthesis over levels

Direct synthesis approach

> Much of the iterative nature can be removed from the planning process, if a synthesis approach is adopted. The essential difference between the analysis-based and synthesis-based procedures is in the level of abstraction adopted in the computations. (The terminology 'level of abstraction' is used in the sense relating to the quantity of a priori data assumed.) Analysis-based techniques impose a total project configuration ab initio, while the plan emerges from any given level of abstraction as a natural consequence of the direct synthetic treatment. Presumably the extreme generality that may be attained in the direct case would involve little or no a priori knowledge of the emerging project. However, for a solution of practical significance, certain leading properties of the project are best assumed. The choice of abstraction level on which the planner chooses to work would be a balance between his/her expertise-based judgements and desired computation load. A synthesis-type treatment can only proceed where certain of the project properties remain free and adjustable. A synthesis-type format to planning is the fundamental approach.
>
> (Carmichael, 2004)

The synthesis approach selects the most appropriate controls from the range of possible controls. The selection is determined by the objectives and constraints.

Within a hierarchy of levels, a problem on one level is nested within the next higher level problem.

Models

When modelling projects, as with most systems, there exists an interdependence among levels. No isolated levels exist. A model at any level contains lower level information, that is, lower level information transfers to higher levels.

The following modelling applies to stripped-back projects, as defined earlier.

Constituent model

The constituent level is the basic behaviour level of a resource (person or piece of equipment).

The model at the constituent level is the resource equivalent to a material technologist's constitutive relationship. Each resource will have a different 'constitutive relationship', as will a resource performing under different conditions. For definiteness, the term 'production' is used here, meaning some measure of work done. A linear example in terms of production is shown in Figure 9.1; nonlinear and integer relationships are possible. The measurement unit of production can be various, for example m^3, m or item.

Schematically, Figure 9.2 applies.

For each resource type,

$$\text{Resource production} = \text{Resource production rate} \times \text{Time period}$$

and

$$\text{Resource unit production} = \text{Resource production rate} \times 1 = \Pi$$

where Π is the resource production rate.

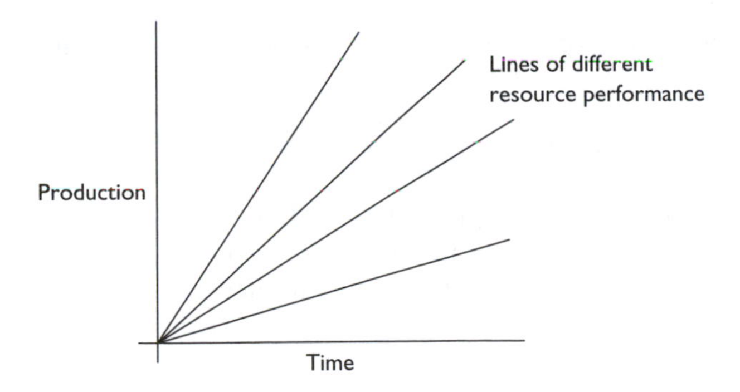

Figure 9.1 Example 'constitutive relationships' for a resource. The slope of a line is a (resource) production rate.

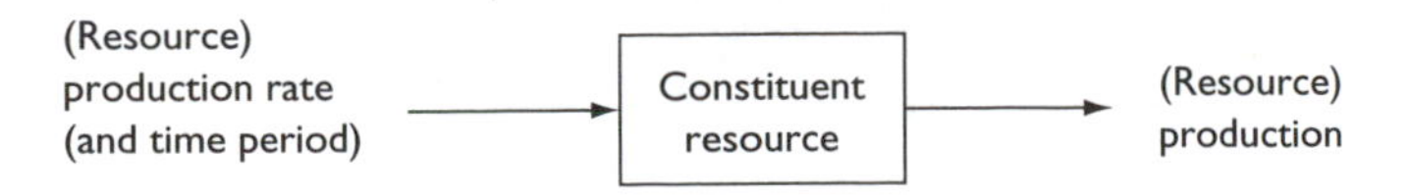

Figure 9.2 Inflow–outflow relationship at the constituent level.

The control at the constituent level is Π for each resource type; resource behaviour is given by its production.

Element model

The inflow–outflow relationship of an element of an activity, in *production* terms, might be represented as in Figure 9.3. An element represents 1 unit of time, for example 1 day or 1 hour.

For each resource type,

$$\begin{aligned}\text{Element production} &= \text{Element resource number} \times \text{Resource unit} \\ &\quad \text{production} \\ &= \rho\Pi\end{aligned}$$

where, ρ is the resource number (quantity), $\rho = 1, 2, \ldots$

If the different resource types have the same unit of measurement for production, the element productions for all resource types can be added to give a total element production.

Figure 9.4 shows the influence of resource numbers (quantities) on production.

In *resource usage* terms, an element uses resource numbers of ρ (for each resource type).

The controls at the element level are $\rho\Pi$ and ρ for each resource type; the behaviour is given by element production and resource number (quantity) for each resource type.

Resource numbers and type can be measured or expressed in money terms.

Activity model

An activity of duration E (for example E days or E hours) is composed of E elements each of duration 1 (for example 1 day or 1 hour).

In terms of *production*, individual elements as in Figure 9.5 combine to give an activity as in Figure 9.6. Typically, the inflow from, and outflow

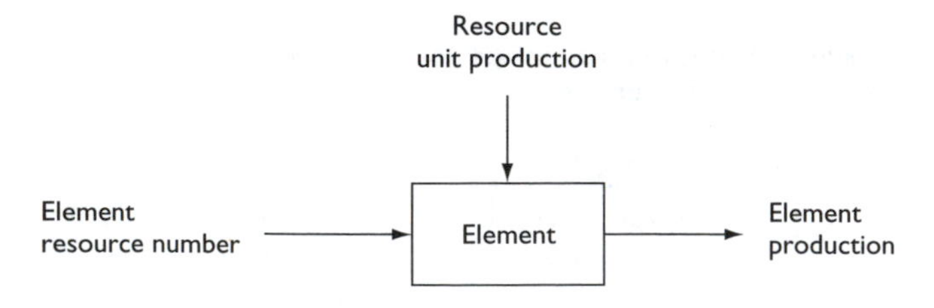

Figure 9.3 Element of an activity; for each resource type.

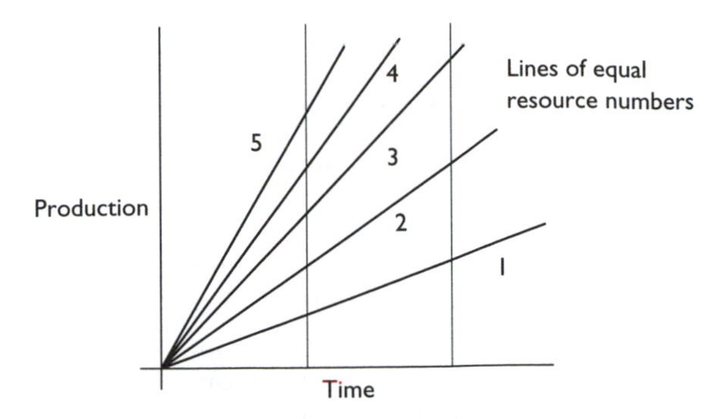

Figure 9.4 Example resource production rate plot for different resource numbers.

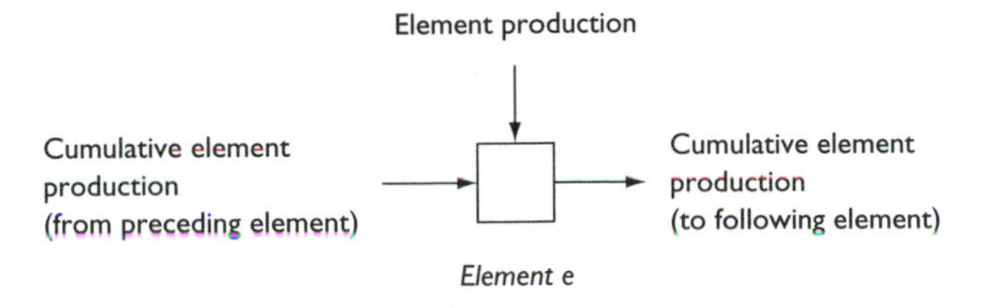

Figure 9.5 Individual element with element interaction.

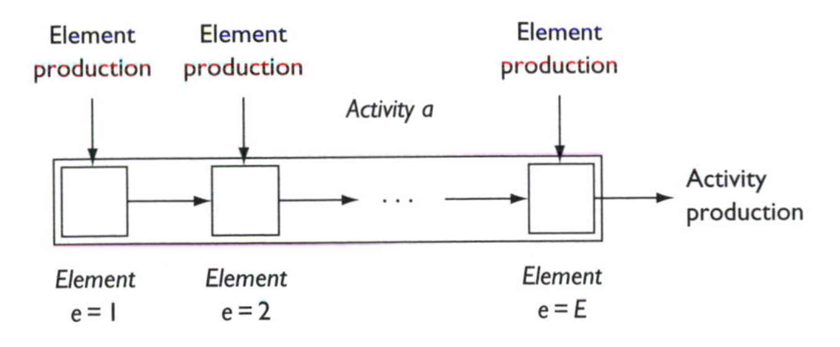

Figure 9.6 Activity; for each resource type.

to, other elements will be cumulative element production, and the control will be element production.

In Figure 9.6, the arrows between elements represent cumulative production.

In Figure 9.5, the element interaction gives, for each resource type,

$$\begin{aligned}&\text{Cumulative element production outflow}\\&\quad= \text{Element production} + \text{Cumulative element production inflow}\\&\quad= \rho + \text{Cumulative element production inflow}\end{aligned}$$

In Figure 9.6, the activity inflow–outflow relationship is, for each resource type,

$$\text{Activity production} = \text{Sum of element productions}$$
$$= \sum_{\text{All elements}} \text{Element productions} = \sum_{\text{All elements}} \rho\Pi$$

In terms of *resource usage*, individual elements as in Figure 9.7 combine to give an activity as in Figure 9.8. Typically, the inflow from, and outflow to other elements will be cumulative element resource numbers, and the control will be element resources (number and type). Resource numbers can vary from element to element, to suit the activity; there is not a requirement for uniform resource numbers.

In Figure 9.8, the arrows between elements represent cumulative element resource numbers.

In Figure 9.7, the element interaction gives, for each resource type,

$$\text{Cumulative element resource number outflow} = \text{Element resource number} + \text{Cumulative element resource number inflow}$$
$$= \rho + \text{Cumulative element resource number inflow}$$

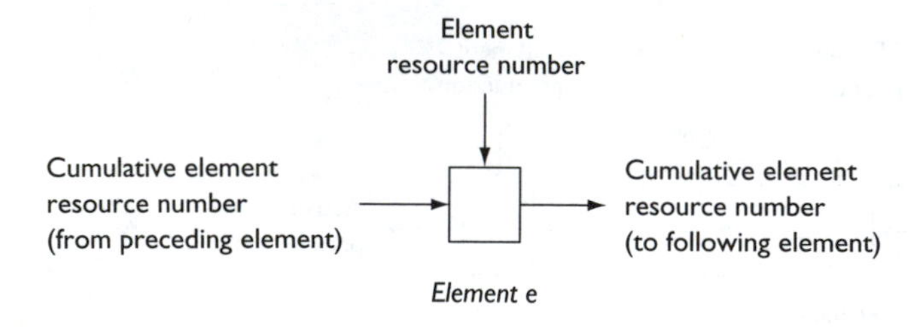

Figure 9.7 Individual element with element interaction.

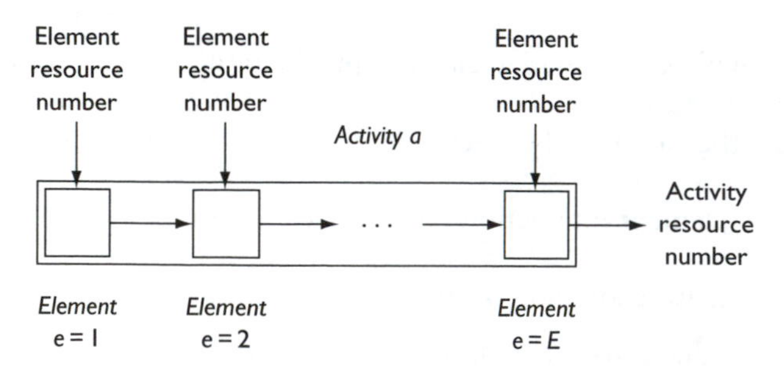

Figure 9.8 Activity composed of elements; for each resource type.

In Figure 9.8, the activity inflow–outflow relationship is, for each resource type,

$$\begin{aligned}\text{Activity resource number} &= \text{Sum of element resource numbers}\\ &= \sum_{\text{All elements}} \text{Element resource numbers}\\ &= \sum_{\text{All elements}} \rho\end{aligned}$$

The controls at the activity level are $\sum_{\text{All elements}} \rho\Pi$ and $\sum_{\text{All elements}} \rho$ for each resource type; the behaviour is given by activity production and activity resource number (quantity), for each resource type.

Resource numbers and type can be measured or expressed in money terms. For the special issue of reducing the duration to perform (compress) an activity, the total resource usage may increase, and the cost may increase as in Figure 9.9.

Project model

Overlapping relationships between activities may be taken care of through using subactivities which are themselves activities. In aggregated form, Figures 9.6 and 9.8, together with interaction between activities, convert to Figures 9.10(a,b) respectively.

For each resource type, for each activity, these can be expressed as, for a *production* focus,

$$\text{Cumulative activity production outflow} = \sum_{\text{All elements}} \text{Element productions} + \text{Cumulative activity production inflow}$$

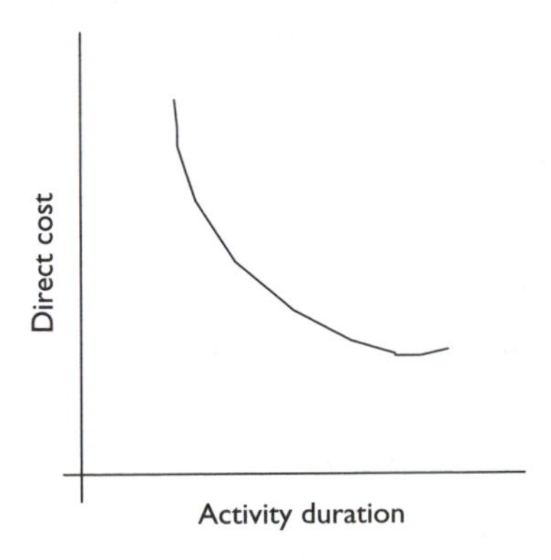

Figure 9.9 Changing activity cost with activity duration.

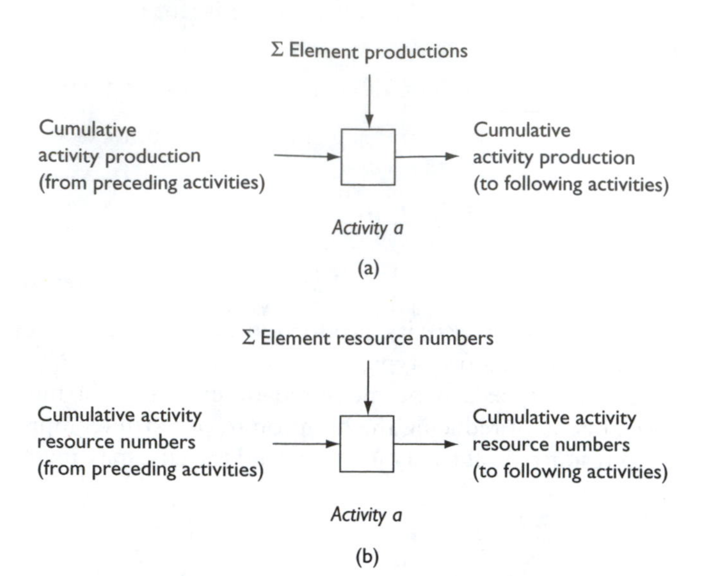

Figure 9.10 Aggregated activity representation. (a) Production focus; (b) Resource usage focus.

For each resource type, for each activity, for a *resource usage* focus,

$$\text{Cumulative activity resource numbers outflow} = \sum_{\text{All elements}} \text{Element resource numbers} + \text{Cumulative activity resource numbers inflow}$$

Activities can be assembled to give a project (or . . . , work package, subsubproject or subproject). For example, consider Figure 9.11.

Activity inflow and outflow will reflect the project network connections, but redundancy of resource and production information through parallel network paths has to be acknowledged. That is, for each activity,

$$\text{Activity inflow} = \text{Other (preceding) activities outflow}$$
$$\text{(ignoring any redundant accounting)}$$
$$\text{Activity outflow} = \text{Activity inflow} + \sum_{\text{All elements}} \text{Element controls}$$

The critical path method equations provide the information on the redundancy, and hence constitute part of the model at the project level.

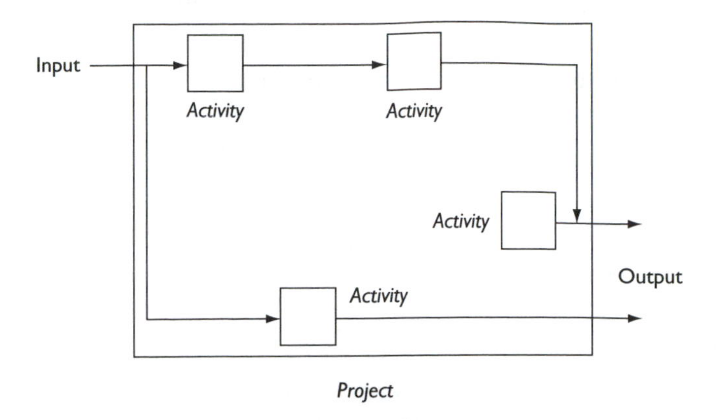

Figure 9.11 Example project composed of activities (not to scale); activity-on-node version.

From a *production* viewpoint, the control will be production related. The arrows between activities (in an activity-on-node context) represent cumulative activity production.

From a *resource usage* viewpoint, the control will be activity resources (number and type). The arrows between activities (in an activity-on-node context) represent cumulative activity resources.

The inflow to and outflow from a project (at the start and end of the project respectively) will be cumulative project resources and cumulative project production. For different project level controls (work methods including activity interaction), different forms of Figure 9.11 will result.

In Figure 9.11, the project inflow–outflow relationship is, for each resource type,

$$\begin{aligned}\text{Project production} &= \text{Sum of activity productions}\\ &= \sum_{\text{All activities}} \sum_{\text{All elements}} \text{Element productions}\\ &= \sum_{\text{All activities}} \sum_{\text{All elements}} \rho\Pi\end{aligned}$$

$$\begin{aligned}\text{Project resource numbers} &= \text{Sum of activity resource numbers}\\ &= \sum_{\text{All activities}} \sum_{\text{All elements}} \text{Element resource numbers}\\ &= \sum_{\text{All activities}} \sum_{\text{All elements}} \rho\end{aligned}$$

The controls at the project level are work method (including sequencing), and $\sum_{\text{All activities}} \sum_{\text{All elements}} \rho\Pi$ and $\sum_{\text{All activities}} \sum_{\text{All elements}} \rho$, for each resource type; the behaviour is given by project production and project resource numbers (quantity), for each resource type.

Resource numbers and type can be measured or expressed in money terms.

Level planning problems

Introduction

A summary of the various level (constituent, element, activity and project) problems together with the relevant variables follows.

Constituent level

The constituent level is the basic behaviour level of a resource (person or piece of equipment).

Control
The control at the constituent level is the resource production rate Π for each resource type.

Output, state
The resource performance is in terms of its production.

Objectives, constraints
Objectives and constraints are resource production or resource production rate connected for different resource types.

Model
See above.

Problem outcome
Resource production rates and resource production. Commonly, these are folded up to the next level.

Element level

An element is a single time unit component of an activity.

Control
The controls at the element level are $\rho\Pi$ and the resource number (quantity) ρ for each resource type.

Output, state
Element performance, for each resource type, is indicated by the element production and the resource number (quantity).

Objectives, constraints
Objectives and constraints will relate to element production, and element resource number and type (for example, resource availability and applicability).

Model
See above.

Problem outcome
Element production, resource numbers and resource types are commonly folded up to the next level.

Resource numbers and type can be measured or expressed in money terms.

Activity level

An activity of duration E is composed of E elements.

Control
The controls at the activity level are $\sum_{\text{All elements}} \rho\Pi$ and $\sum_{\text{All elements}} \rho$ for each resource type and activity.

Output, state
The activity performance, for each resource type, is indicated by activity production and the activity resource number (quantity).

Objectives, constraints
Objectives and constraints will relate to activity production, and activity resource number and type.

There may be resource availability and applicability issues. Particular resource usages may apply, for example those relating to particular work practices of the work crews and subcontractors.

Model
See above.

Problem outcomes
The resource (or money) usage within activities, and total activity resource usage and activity production.

Resource numbers and type can be measured or expressed in money terms.

Project level

The project level also includes subproject, sub-subproject and similar levels, and represents a collection of connected activities.

Control
The controls at the project level are work method (including sequencing), and $\sum_{\text{All activities}} \sum_{\text{All elements}} \rho\Pi$ and $\sum_{\text{All activities}} \sum_{\text{All elements}} \rho$, for each resource type.

Output, state
Project performance, for each resource type, is indicated by project production and project resource number (quantity).

Objectives, constraints
Objectives and constraints will relate to project production, project resource number and type, and work method.

The total work will have to conform to the (project) scope.

Owner-specified objectives and constraints will carry down to lower levels.

Model
See above.

Problem outcomes
The work method, and resource (or money) usage within the project, and total project resource usage and project production.

Resource numbers and type can be measured or expressed in money terms.

General comment

Generally the higher the level, the harder the problem, but the more useful the information coming out of the problem.

A common approach (subproblem) is to fix the work method, and only allow choice in resource usage and resource production rates. That is, only the lower level problems are solved. The possibility of multiple work methods or networks (involving activity sequencing) is commonly not left open, but rather assumed known (unless an iterative-analysis approach is adopted). The synthesis problem of determining a preferred network from a class of networks is too difficult. It is analogous to Michell frame-structures in structural engineering (Carmichael, 1981), for which only a few special results are known.

Upon establishing the controls at all levels, then additional outcomes include:

- The timing of the activities. This includes the start and finish times of activities, allowing a bar chart or generally a program to be drawn
- Project milestones and target dates
- Critical project activities
- Expenditure, budget, cash flow (additionally requiring estimates of income)
- Resource plots
- Delivery requirements.

Multilevel optimisation thinking

Introduction

Systems may be decomposed to subsystems with interaction. Likewise, optimisation at the system level can be decomposed to optimisations at the subsystem level together with interaction. Lower level problems could be expected to be easier to solve, though the coupling between subproblems introduces iterations into the computations.

The lower level is generally referred to as the infimal or first level, while the upper level is referred to as the supremal or second level (Figure 9.12). The upper level coordinates the lower levels, such that the solution to the original problem is obtained. Extensions to further levels are possible, although for practical project planning reasons decomposition would possibly not go below activity level.

Decomposition of the system level problem can be done based on a physically meaningful decomposition or non-physically meaningful decomposition of the system, though the former would be more attractive for planners and could be expected to provide less complicated solutions. For

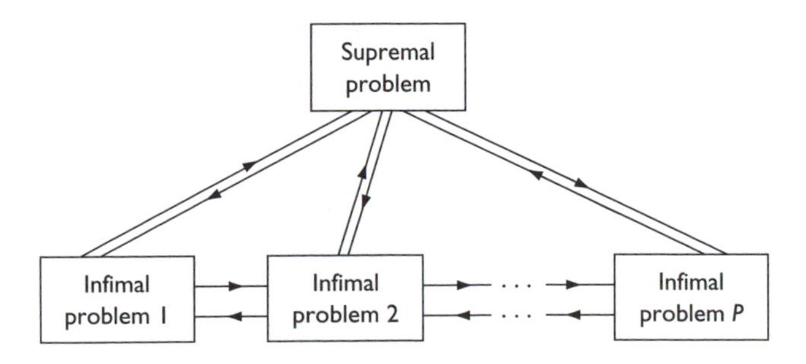

Figure 9.12 Problem decomposition and interaction.

planning problems, the most obvious project decomposition choices would be according to conventional work breakdown structure thinking:

- Phase (chronological); time period phasing
- Work type or nature; resource groupings
- Work parcels; the way the work is to be carried out
- Contracts
- Region, location or geographical area
- Organisational breakdown
- (Existing) cost codes, cost centres, cost headings; project cost and data summary reports
- The nature of the planning network; activity groupings
- Function.

But other choices are possible. The decomposition chosen is at the discretion of the planner.

Phase or stage treatment of planning is also considered in Chapter 8. There the project duration is broken into subintervals. The subintervals become the subprojects, and may have some physical meaning, or may be chosen for some other, perhaps computational, reason.

The solution of a multilevel optimisation problem requires *decomposition* and *coordination*. Decomposition implies taking the project level problem components – model, objectives and constraints – and breaking these down to models, objectives and constraints at the subproject level. Interaction variables are introduced in order to separate the subproblems. Coordination of the subproblem solutions ensures that the solution to the original project level problem is obtained. Coordination requires a repeated two-way iterative transfer between the levels, adjusting the coordination until a desired solution is obtained.

Generally the solution follows Figure 9.13 or Figure 9.14. A number of choices are available for selecting the coordination variables.

The networks shown in Figure 9.15 are used in examples in the following.

Decomposition according to the system model

Chapter 8 gives an example staging of the networks shown in Figure 9.15. There system equations of the following form arise,

$$\begin{bmatrix} x3 \\ x5 \\ x6 \\ x7 \\ x10 \\ x12 \end{bmatrix} = \begin{bmatrix} x1 \\ x3 \\ x3 \\ x3 \\ x6 + x7 - x3 \\ x10 \end{bmatrix} + \begin{bmatrix} u1 + u2 \\ u4 \\ u3 + u5 \\ u3 + u4 \\ u7 + u9 + u10 + u8 - u3 \\ u11 + u6 + u12 \end{bmatrix}$$

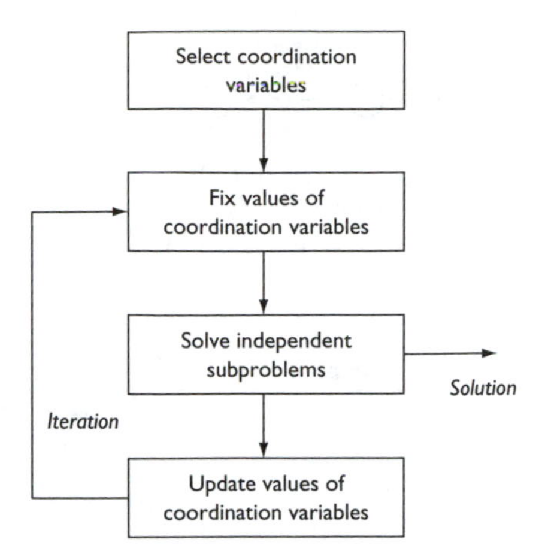

Figure 9.13 Iterative solution to the multilevel optimisation problem.

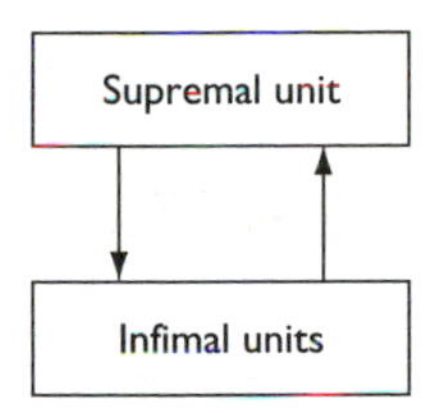

Figure 9.14 Iteration between levels.

where x*j* is the output/state following activity (*i*, *j*), u(*i*, *j*) is the control for activity (*i*, *j*), and where the following has been introduced to simplify the notation,

$$
\begin{aligned}
u1 &= u(1,2)\\
u2 &= u(2,3)\\
u3 &= u(3,4)\\
u4 &= u(3,5)\\
u5 &= u(4,6)\\
u6 &= u(5,11)\\
u7 &= u(6,8)\\
u8 &= u(7,9)\\
u9 &= u(8,9)\\
u10 &= u(9,10)\\
u11 &= u(10,11)\\
u12 &= u(11,12)
\end{aligned}
$$

u(4,7) and u(5,7) relate to dummy activities and have been omitted.

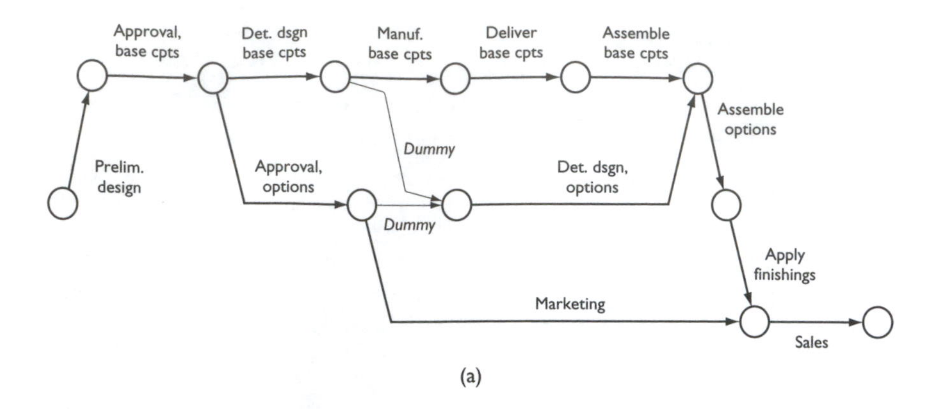

(a)

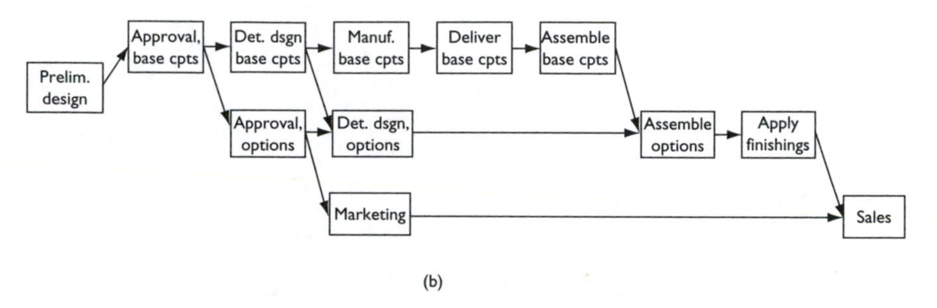

(b)

Figure 9.15 Example network diagrams. (a) Activity-on-link diagram; (b) Activity-on-node diagram. (Activity names have been abbreviated because of space limitations – work items are implied.)

The state represents cumulative resources and cumulative production. The control represents the chosen resource numbers and production.

Assume a minimum duration (maximum production) objective of

$$\min \mathrm{J} = -\sum_{i=1}^{12} \mathrm{u}i$$

(where only the ui components related to production are considered).

And a constraint related to resource usage,

$$\mathrm{u}i \le \mathrm{U}_i \quad i = 1, 2, \ldots, 12$$

where U_i, $i = 1, 2, \ldots, 12$, are set resource numbers (and where only the ui components related to resource usage are considered).

Other constraints would generally be present for a meaningful solution. As well, the objective could be expected to be more involved. However, the objective and constraint given here are sufficient for the present demonstration purposes.

The decomposition selected for this example partitions the system equations – in effect a horizontal line is drawn two-thirds the way down through the above system equations. This corresponds to the division between stages 1 and 2 in Figure 9.16.

For this particular decomposition, to separate the subproblems, introduce the interaction variables.

$$\pi^1 = x6$$
$$\pi^2 = x7$$
$$\pi^3 = x3$$
$$\pi^4 = x3$$

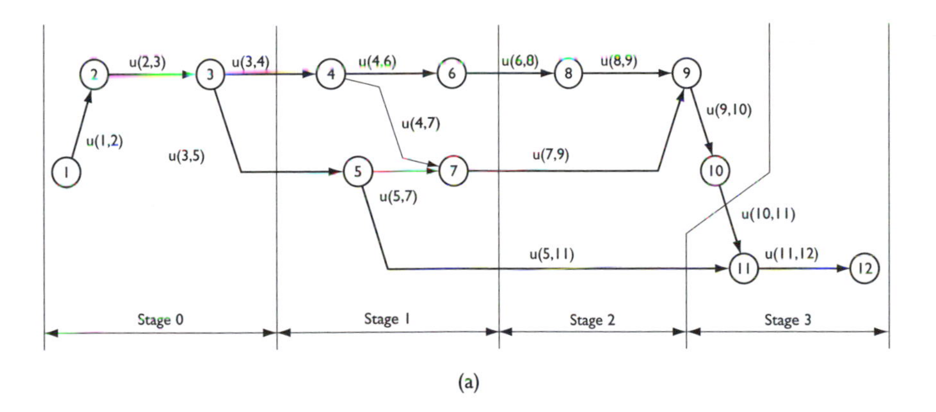

(a)

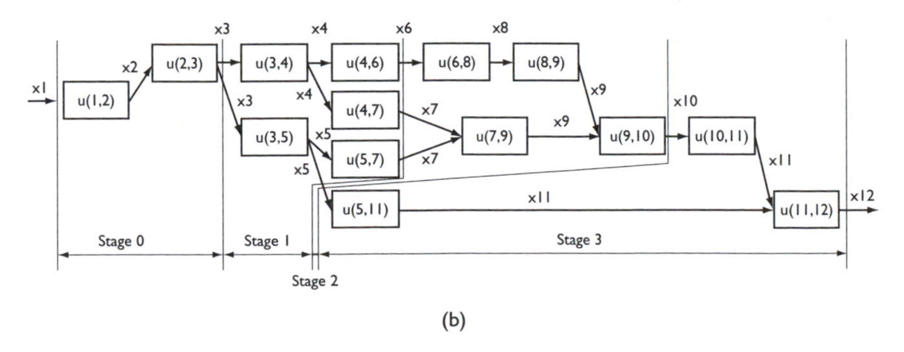

(b)

Figure 9.16 Example network diagrams. (a) Activity-on-link diagram; (b) Activity-on-node diagram.

The objective becomes

$$\min \mathrm{J} = -\sum_{i=1}^{12} \mathrm{u}i + \beta^1(\pi^1 - \mathrm{x6}) + \beta^2(\pi^2 - \mathrm{x7}) + \beta^3(\pi^3 - \mathrm{x3}) + \beta^4(\pi^4 - \mathrm{u3})$$

where $\beta^i, i = 1, 2, \ldots, 4$ are Lagrange multipliers.

Subproblem one

$$\min \mathrm{J}^1 = \min - \sum_{i=1}^{5} \mathrm{u}i + \beta^1\mathrm{x6} + \beta^2\mathrm{x7} + \beta^3\mathrm{x3} + \beta^4\mathrm{u3}$$

subject to

$$\begin{bmatrix} \mathrm{x3} \\ \mathrm{x5} \\ \mathrm{x6} \\ \mathrm{x7} \end{bmatrix} = \begin{bmatrix} \mathrm{x1} \\ \mathrm{x3} \\ \mathrm{x3} \\ \mathrm{x3} \end{bmatrix} + \begin{bmatrix} \mathrm{u1} + \mathrm{u2} \\ \mathrm{u4} \\ \mathrm{u3} + \mathrm{u5} \\ \mathrm{u3} + \mathrm{u4} \end{bmatrix}$$

$$\mathrm{u}i \leq \mathrm{U}_i \quad i = 1, 2, \ldots, 5$$

Subproblem two

$$\min \mathrm{J}^2 = \min - \sum_{i=6}^{12} \mathrm{u}i + \beta^1\pi^1 + \beta^2\pi^2 + \beta^3\pi^3 + \beta^4\pi^4$$

subject to

$$\begin{bmatrix} \mathrm{x10} \\ \mathrm{x12} \end{bmatrix} = \begin{bmatrix} \pi^1 + \pi^2 - \pi^3 \\ \mathrm{x10} \end{bmatrix} + \begin{bmatrix} \mathrm{u7} + \mathrm{u9} + \mathrm{u10} + \mathrm{u8} - \pi^4 \\ \mathrm{u11} + \mathrm{u6} + \mathrm{u12} \end{bmatrix}$$

$$\mathrm{u}i \leq \mathrm{U}_i \quad i = 6, 7, \ldots, 12$$

Solution

One possible solution fixes the values for the Lagrange multipliers $\beta^i, i = 1, 2, \ldots, 4$, and solves the suboptimisation problems for the xs, us and πs. The values of $\beta^i, i = 1, 2, \ldots, 4$ are then updated such that x6 approaches π^1, x7 approaches π^2, x3 approaches π^3 and u3 approaches π^4. And the process repeats.

Other solution approaches are possible, as are other decompositions of the original problem.

Decomposition according to hierarchy

As an example of a problem involving hierarchical decomposition assume, as before, a minimum duration (maximum resource usage/expenditure) objective of

$$\min \mathrm{J} = -\sum_{i=1}^{12} \mathrm{u}i$$

and a constraint related to resource usage,

$$\mathrm{u}i \le \mathrm{U}_i \quad i = 1, 2, \ldots, 12$$

where U_i, $i = 1, 2, \ldots, 12$, are set resource numbers, together with the example networks of Figures 9.15 and 9.17.

The notation used is the same as in the previous example.

As before, other constraints would generally be present for a meaningful solution. As well, the objective could be expected to be more involved. However, the objective and constraint given here are sufficient for the present demonstration purposes.

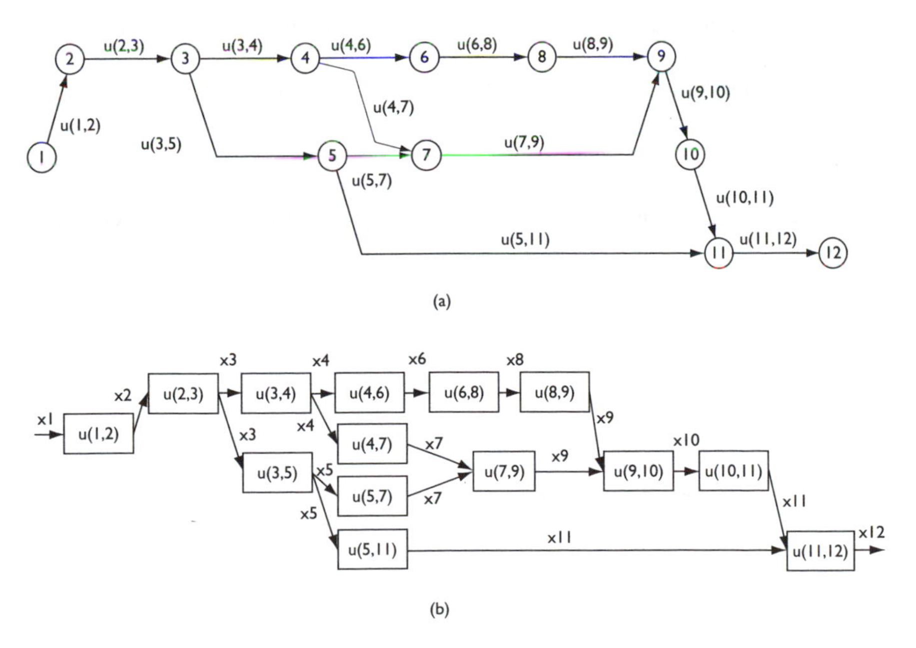

Figure 9.17 Example network diagrams. (a) Activity-on-link diagram; (b) Activity-on-node diagram.

At the project level, the system equations can be written using the notation of Figure 9.17 as,

$$\begin{bmatrix} x2 \\ x3 \\ x4 \\ x5 \\ x6 \\ x7 \\ x8 \\ x9 \\ x10 \\ x11 \\ x12 \end{bmatrix} = \begin{bmatrix} x1 + u(1,2) \\ x2 + u(2,3) \\ x3 + u(3,4) \\ x3 + u(3,5) \\ x4 + u(4,6) \\ x4 + u(4,7) \quad + x5 + u(5,7) \\ x6 + u(6,8) \\ x8 + u(8,9) \quad + x7 + u(7,9) \\ x9 + u(9,10) \\ x10 + u(10,11) + x5 + u(5,11) \\ x11 + u(11,12) \end{bmatrix} - \begin{bmatrix} \\ \\ \\ \\ \\ x3 \\ \\ x4 \\ \\ x5 \\ \\ \end{bmatrix}$$

with x1 given. Or a matrix equation in 12 rows could be given if x1 is included.

Here xj is the output/state following activity (i, j), and $u(i, j)$ is the control for activity (i, j).

In summarised form,

$$X = M - B$$

where X, M and B are column vectors of 11 rows, with components xi, mi and bi respectively. $i = 2, 3, \ldots, 12$; or $i = 1, 2, \ldots, 12$.

This can be built up from the activity level as shown in the following.

Activity model

(a) At the activity level (Figure 9.18), the system equation for each activity is

$$x'j = x^*i + u(i, j)$$

where

x^*i	the state following node i
$x'j$	the state leading into node j
$u(i, j)$	control for activity (i, j).

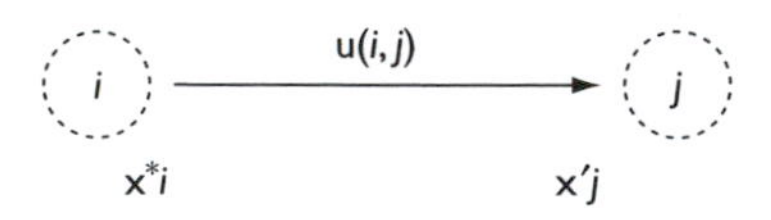

Figure 9.18 Activity notation; activity-on-link version.

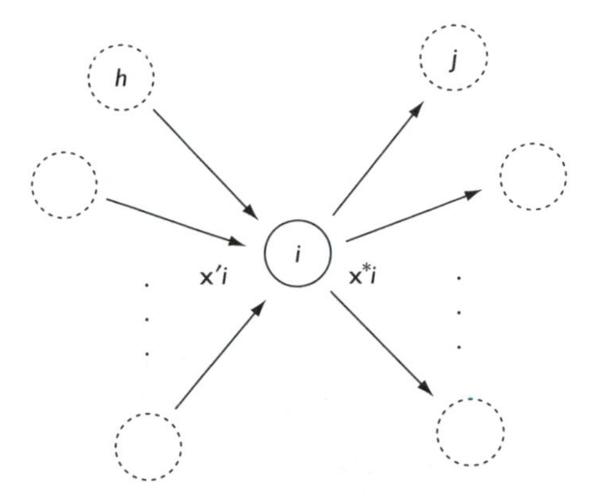

Figure 9.19 Continuity at a node; activity-on-link version.

Activity interaction

(b) *Continuity at a node* (Figure 9.19),

$$x^{*}i = \sum_{\substack{\text{All preceding} \\ \text{activities}(h,i)}} x'i$$

(c) *Activity connectivity* that accommodates redundant paths through a network may be expressed by

$$x''j = -bj$$

where

$x''j$ correction to the state leading into node j because of redundancies in the network.

Project model

The project model is obtained by combining (a), (b) and (c) to give

$$xj = \sum_{\substack{\text{All preceding} \\ \text{activities}}} [xi + u(i, j)] - bj$$

or

$$X = M - B$$

where

$$mj = \sum_{\substack{\text{All preceding}\\ \text{activities}}} [xi + u(i, j)]$$

and xj is the state following node j.

Optimisation

To solve the project level optimisation problem as a collection of activity level subproblems, the activity interaction, the interaction equations (b) and (c) are adjoined to the original objective,

$$J' = J + \sum_{i=1}^{12} \beta 1_i [x^*i - \sum_{\substack{\text{all}\\ (h,i)}} x'i] + \sum_{j=1}^{12} \beta 2_j [x''j + bj]$$

where $\beta 1$ and $\beta 2$ are Lagrange multipliers.

Consider, for example, the subproblem associated with activity (11,12),

$$\min J' = -u(11,12) + \beta 1_{11} x11 - \beta 1_{12} x'12 + \beta 2_{11} x''11$$

subject to (using the simpler notation of the previous chapter),

$$x12 = x11 + u(11,12)$$

$$u(11, 12) \leq U_{11,12}$$

Other formulations can be given.

Chapter 10

Selected topic

Project compression

Introduction

The project compression problem is a subproblem of the planning problem. Typically, the project level control (work method) is held fixed and only lower level controls are optimised.

A number of approaches have been suggested in the literature for solving the project compression problem as an optimisation problem, typically a linear optimisation or linear programming problem. The optimisation approaches do not appear to have found favour in commercial project planning programs or with practitioners. Some reasons suggested for this are:

- Most commercial project planning packages appear to be based on algorithms that reflect the particular characteristics of networks. In this way the algorithm can be written in general terms and applied to all networks. A linear programming approach, on the other hand, requires input peculiar to each network – the network logic of activity precedence is interpreted as constraints in the linear programming approach. It is conceivable that a transformation could be written in general terms giving the network logic as constraints.
- Knowledge of linear programming and how to formulate problems as optimisation problems is not widespread. As well, linear programming packages tend not to be readily available, although some spreadsheets do include this option.

Nevertheless, it is informative to see the linear programming formulation to the project compression problem.

The development below follows Carmichael (1996).

Cost–duration relationships

A cost–duration relationship or cost function such as Figure 10.1 is typically assumed.

The equation of the cost–duration relationship is

$$c_i = C_i - s_i y_i \quad YC_i \leq y_i \leq YN_i$$

where

y_i	duration of activity i
c_i	direct cost of activity i
s_i	cost slope for activity i
C_i	intercept on the cost axis for activity i
i	activity number; $i = 1, 2, \ldots, N$
N	total number of activities
CN_i	normal cost
CC_i	crash cost
YN_i	normal duration
YC_i	crash duration.

Cost functions other than linear

Many activities could be expected to have convex cost functions or cost functions that are not linear (Figure 10.2). For convenience, a linear relationship might be assumed going between the normal and crash points. Alternatively, a piecewise linear approximation might be assumed.

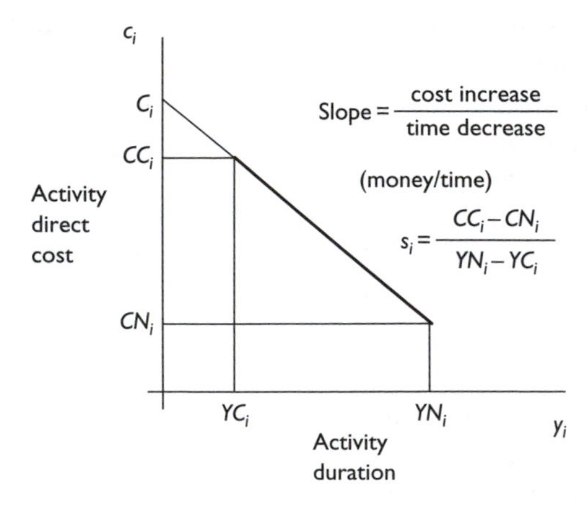

Figure 10.1 Linear cost–duration relationship for activity i; $i = 1, 2, \ldots, N$.

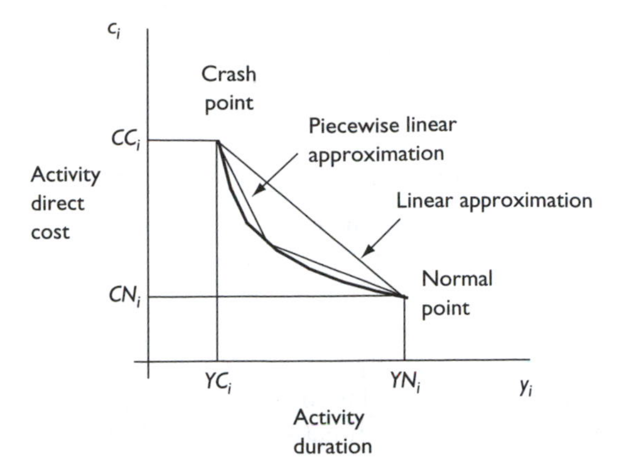

Figure 10.2 Convex cost function for an activity.

The accuracy of any approximation is not an issue here because the accuracy with which the normal and crash points and all points in between is known is uncertain.

Where the cost–duration relationship is integer in form, then continuous approximations might be assumed, or integer (linear) programming used. Integer-based optimisation techniques do not seem as popular as other optimisation techniques.

The optimisation problem

Any optimisation problem typically involves three components; an objective function, a system model and constraints. For the project compression problem these are as follows.

Objective function

The objective function is usually one of cost and a minimum is sought. The total (direct) cost is the sum of the activity costs.

$$\min J = \sum_{i=1}^{N} C_i - s_i y_i$$

or equivalently

$$\max J = \sum_{i=1}^{N} s_i y_i$$

Here minimisation has been replaced by maximisation and a sign reversal. The constants have been removed because these contribute nothing to establishing the optimum activity durations (but they do contribute to the total [direct] cost).

Bonus for early completion, penalty for late completion

For a variable project end date, the duration by which the end date is less than or greater than a nominated duration, T_{project}, is

$$|T_{\text{project}} - \text{LFT}_N|$$

where LFT_N is the latest finish time of the last project activity.

Any bonus, B, or penalty, P, will be a function of this. Either B or P applies but not both.

The objective function becomes

$$\max \text{J} = \sum_{i=1}^{N} s_i y_i + B - P$$

B and P could also be used like penalty functions to force a solution close to a desired project completion date, in the absence of any project bonuses or penalties.

Constraints

A number of constraints, restricting the choice of the optimum solution, apply.

Budget constraints restrict the total amount of money that can be spent to a fixed amount. That is,

$$\sum_{i-1}^{N} C_i - s_i y_i \leq C_{\text{project}}$$

where C_{project} is the maximum amount of money to be spent on the project. Where C_{project} is not known, it may be left as something to be determined from the optimisation. That is, this constraint will not apply.

Activity duration ranges give the ranges over which the activity cost–duration relationships apply and possibly also the workable durations of activities.

$$YC_i \leq y_i \leq YN_i \quad i = 1, 2, \ldots, N$$

Non-negativity constraints, although seemingly obvious, are added to ensure a sensible solution. That is, all variables are constrained to be non-negative.

$$y_i \geq 0 \quad i = 1, 2, \ldots, N$$

As well, in some formulations, event times may also be variables.

$$\mathrm{EST}_i \geq 0$$
$$\mathrm{LST}_i \geq 0$$
$$\mathrm{EFT}_i \geq 0$$
$$\mathrm{LFT}_i \geq 0$$
$$\mathrm{TF}_i \geq 0 \qquad i = 1, 2, \ldots, N$$

Here,

EST_i	earliest start time of activity i
LST_i	latest start time of activity i
EFT_i	earliest finish time of activity i
LFT_i	latest finish time of activity i
TF_i	total float of activity i

For activity-on-link diagrams, these could be expressed directly as event times rather than as start or finish times of activities.

Milestone constraints. For various reasons, there may be events that have to occur before/after some specified time or date.

$$\mathrm{LFT}_i \leq \text{some value} \quad \text{for some } i$$
$$\mathrm{EST}_i \geq \text{some value} \quad \text{for some } i$$

For example, it may be desirable to set a duration above which the project duration could not go,

$$\mathrm{LFT}_N \leq \text{some value}$$

For activity-on-link diagrams, these constraints could be expressed as event time constraints.

Integer value constraints. Generally it could be expected that activity durations will be required to be integer valued though this need not be the case.

This implies using an integer-based optimisation technique or using approximations of some form.

Subprojects. Where the project consists of subprojects or work packages, each with their own requirements, similar constraints to all the above will apply. A reformulation in terms of subprojects or work packages adds nothing new to the above generalisations; the total number of constraints is the sum of the subproject constraints together with something relating the subproject costs to the total (direct) cost.

A subproject is a project. Any formulation for projects is straightforwardly adaptable to subprojects.

System model

The behaviour of the network comes from the precedence relationships as defined by the network.

That is, the system model is the usual equations that give the earliest start times, latest start times, earliest finish times, latest finish times and total floats for the activities, through typically forward pass and backward pass treatments.

Where overlapping relationships apply, the equations are slightly more involved than for the non-overlapping relationships case.

Solution

Published approaches to this problem spell out all these constraints, one by one, even the network logic which is interpreted as multiple constraints. For a network involving many activities, this can be a lengthy process. An optimisation package is then used to find a solution.

Example

Consider the problem posed in Deckro *et al.* (1992). The network is shown in Figure 10.3. The cost–duration data are shown in Table 10.1.

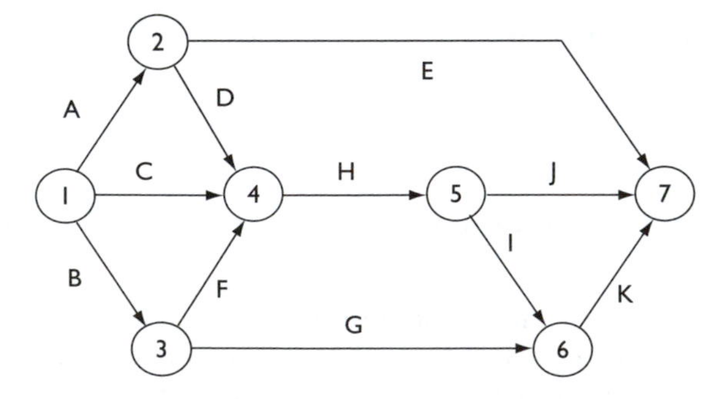

Figure 10.3 Network diagram for example. Activity-on-link diagram.

Table 10.1 Cost–duration data for example

Activity	*Nodes (begin, end)*	*Activity duration, lower limit*	*Activity duration, upper limit*	*Cost function slope ($/day)*	*Cost function intercept ($)*
A	(1,2)	7	10	5	2131
B	(1,3)	6	8	7	3274
C	(1,4)	6	13	6	9341
D	(2,4)	6	6	–	4070
E	(2,7)	20	28	9	2010
F	(3,4)	3	5	8	8519
G	(3,6)	15	23	11	9708
H	(4,5)	3	8	3	434
I	(5,6)	4	9	8	7773
J	(5,7)	6	10	10	9440
K	(6,7)	5	11	6	9542

The project is assumed composed of two subprojects or work packages comprising activities {A, B, C, D, F} and {E, G, H, I, J, K}.

In the following, y stands for an activity duration, while e stands for an event time. Both activity durations and event times are variables in this problem formulation.

Objective function

$$\begin{aligned}\min \mathrm{J} &= (2131 - 5y_{\mathrm{A}}) + (3274 - 7y_{\mathrm{B}}) + (9341 - 6y_{\mathrm{C}}) + (4070) \\ &\quad + (2010 - 9y_{\mathrm{E}}) + (8519 - 8y_{\mathrm{F}}) + (9708 - 11y_{\mathrm{G}}) + (434 - 3y_{\mathrm{H}}) \\ &\quad + (7773 - 8y_{\mathrm{I}}) + (9440 - 10y_{\mathrm{J}}) + (9542 - 6y_{\mathrm{K}}) \\ &= 66\,242 - 5y_{\mathrm{A}} - 7y_{\mathrm{B}} - 6y_{\mathrm{C}} - 9y_{\mathrm{E}} - 8y_{\mathrm{F}} - 11y_{\mathrm{G}} - 3y_{\mathrm{H}} - 8y_{\mathrm{I}} \\ &\quad - 10y_{\mathrm{J}} - 6y_{\mathrm{K}}\end{aligned}$$

or

$$\max \mathrm{J} = 5y_{\mathrm{A}} + 7y_{\mathrm{B}} + 6y_{\mathrm{C}} + 9y_{\mathrm{E}} + 8y_{\mathrm{F}} + 11y_{\mathrm{G}} + 3y_{\mathrm{H}} + 8y_{\mathrm{I}} + 10y_{\mathrm{J}} + 6y_{\mathrm{K}}$$

Constraints

Budget constraints

For an assumed project budget of $65 500,

$$\begin{aligned}&66\,242 - 5y_{\mathrm{A}} - 7y_{\mathrm{B}} - 6y_{\mathrm{C}} - 9y_{\mathrm{E}} - 8y_{\mathrm{F}} - 11y_{\mathrm{G}} - 3y_{\mathrm{H}} - 8y_{\mathrm{I}} - 10y_{\mathrm{J}} - 6y_{\mathrm{K}} \\ &\quad \leq 65\,500\end{aligned}$$

For separate work package budgets of \$27 200 and \$38 300 respectively,

$$(2131 - 5y_A) + (3274 - 7y_B) + (9341 - 6y_C) + (4070) + (8519 - 8y_F) \leq 27\,200$$

$$(2010 - 9y_E) + (9708 - 11y_G) + (434 - 3y_H) + (7773 - 8y_I) + (9440 - 10y_J) + (9542 - 6y_K) \leq 38\,300$$

Due date constraints
For the project to finish by day 31

$$e_7 \leq 31$$

or for the two work packages to finish by days 15 and 31 respectively

$$e_4 \leq 15 \quad \text{and} \quad e_7 \leq 31$$

Activity duration ranges

$$7 \leq y_A \leq 10$$

$$6 \leq y_B \leq 8$$

$$6 \leq y_C \leq 13$$

$$6 \leq y_D \leq 6 \quad \text{or} \quad y_D = 6$$

$$20 \leq y_E \leq 28$$

$$3 \leq y_F \leq 5$$

$$15 \leq y_G \leq 23$$

$$3 \leq y_H \leq 8$$

$$4 \leq y_I \leq 9$$

$$6 \leq y_J \leq 10$$

$$5 \leq y_K \leq 11$$

Bounds on event times

$$e_1 \leq 0$$

$$e_2 \leq 0$$

$$e_3 \leq 0$$

$$e_4 \leq 0$$

$$e_5 \leq 0$$

$$e_6 \leq 0$$

$$e_7 \leq 0$$

These non-negative constraints ensure a sensible solution. The non-negativity constraints on the activity durations can be omitted because they are taken care of in the constraints relating to activity duration ranges.

System model

The behaviour of the network is given by the network's sequencing relationships which are here regarded as additional constraints.

$$e_2 \geq y_A + e_1$$

$$e_3 \geq y_B + e_1$$

$$e_4 \geq y_C + e_1$$

$$e_4 \geq y_D + e_2$$

$$e_4 \geq y_F + e_3$$

$$e_5 \geq y_H + e_4$$

$$e_6 \geq y_I + e_5$$

$$e_6 \geq y_G + e_3$$

$$e_7 \geq y_E + e_2$$

$$e_7 \geq y_J + e_5$$

$$e_7 \geq y_K + e_6$$

Solution

The problem is a linear programming problem in 18 variables (11 activity durations and 7 event times).

The number of constraints equals 33 and is made up as follows:

Budget constraints	2
Due date constraints	2
Activity duration ranges	11
Event times bounds	7
System model	11
	33

A summary of the solution is given in Table 10.2(a, b and c).

Table 10.2 Solution summary of example

Activity	*Nodes (begin, end)*	*Activity duration (days)*	*Activity cost ($)*
A	(1,2)	7	2 096
B	(1,3)	6	3 232
C	(1,4)	13	9 263
D	(2,4)	6	4 070
E	(2,7)	24	1 794
F	(3,4)	5	8 479
G	(3,6)	20	9 488
H	(4,5)	4	422
I	(5,6)	9	7 701
J	(5,7)	10	9 340
K	(6,7)	5	9 512
			65 397

(a)

Node	*Realisation (days)*
1	0
2	7
3	6
4	13
5	31
6	26
7	31

(b)

	Completion duration (days)	Cost ($)	*Amount below budget ($)*
Work package 1	13	27 140	60
Work package 2	31	38 257	43
Project	31	65 397	103

(c)

A compromise approach

An alternative approach is possible.

Rather than solve the problem as a pure optimisation problem, it is considered that the strength and efficiency of the forward pass – backward pass algorithm of network analysis, together with the nice features of optimisation, be employed and the optimisation problem solved in an iterative analysis fashion.

The approach suggested is a comprehensive computer-based method for the optimal cost–duration problem. The approach contains these features:

- Network analysis based on the efficient forward pass – backward pass algorithm.
- An optimisation method based on sequential linearisation which is capable of solving problems involving large networks with many constraints.
- The constraints and objective function may be any continuous nonlinear functions involving activity costs and activity durations.

The approach is illustrated by applying it to an example problem.

Using a straight optimisation approach, the requirement of handling large networks and nonlinear objective functions and constraints, constitutes a dilemma. Also the straight optimisation approach requires a different formulation for activity-on-link diagrams and activity-on-node diagrams. A solution to these dilemmas lays in adopting an iterative-analysis-optimisation approach. Both network types are covered by the approach. The approach is illustrated schematically in Figure 10.4.

Many of the proposed optimisation approaches are specifically for either activity-on-link or activity-on-node diagrams. The iterative-analysis-optimisation approach makes no such assumptions.

All the approaches that use direct optimisation seem to suffer from the fact that the optimisation problem has to be set up for each case separately. This is on top of having to set up the network in the first place. The optimisation problem, for example if linear, requires setting up the left- and right-hand side coefficients of the constraints.

The iterative-analysis-optimisation approach, on the other hand, only requires to set up the network which is repeatedly analysed, on proceeding to the optimum solution.

Within each cycle, the following is carried out:

- A network analysis using the latest compressed values, $y_i^0, i = 1, 2, \ldots, N$, of activity durations. To start the algorithm off, commonly activity normal durations might be assumed.
- Optimisation is then carried out in order to compress the project further, that is, reduce some of the activity durations.
- These new values, $y_i^0, i = 1, 2, \ldots, N$, of the activity durations are then input for the network analysis of the following cycle.

The characteristics of the optimisation problem solved within each cycle are:

- Both constraints and objective function, if nonlinear, are linearised about the latest network configuration, denoted by a superscript ‘0’.

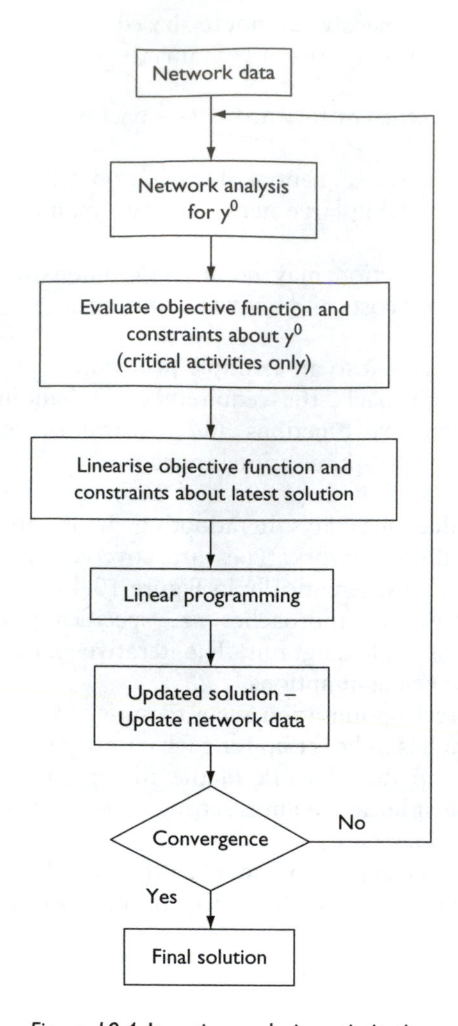

Figure 10.4 Iterative-analysis-optimisation scheme of solution.

A nonlinear function, g(x), may be linearised by using the first order terms of a Taylor series expansion about a solution x^0.

$$g(x) \equiv g(x^0) + (x_1 - x_1^0)\frac{\partial g(x^0)}{\partial x_1} + \ldots + (x_n - x_n^0)\frac{\partial g(x^0)}{\partial x_n} + \text{higher order terms}$$

where $x = [x_1, x_2, \ldots, x_n]^T$.

- Usual activity duration range constraints, budget limit constraints and non-negativity constraints apply.
- The compression is carried out by reducing activity durations one time unit (day, week, ...) by one time unit.

  This implies that the linear approximation to the nonlinear objective function and constraints is reasonable.

  But for this restriction on the compression to apply, step constraints need to be introduced. That is, the activity durations are constrained from reducing by more than one time unit. These step constraints may be written as

$$y_i \geq y_i^0 - 1$$

- Only activities shown to be critical through the network analysis are included as variables in the optimisation problem.
- Milestone constraints are dealt with in the analysis phase, and can be dealt with as the compression proceeds time unit by time unit.

The algorithm has two features which are not of any real consequence but may be considered undesirable by some people:

- In any one cycle the compression may occur in several critical activities simultaneously, provided the constraints are not violated. Purists might suggest that compression should be carried out in order of increasing cost slopes of activities. This algorithm does not rank the cost slopes unless a choice between critical activities is required to avoid violating a constraint.
- Where parallel critical paths exists, compression occurs on all paths, but this compression may be by different amounts on each path. What this translates to is in the analysis on the next cycle, the critical path(s) which was compressed least becomes critical and the other paths transfer from being critical to non critical. The optimisation phase then corrects for this and returns the network to a parallel critical path situation. That is, there may be a situation of over-compressing or overshooting followed by a catching up or correction in the next cycle. This over-compressing can always be checked and allowed for because it gives cost increases not commensurate with the project duration decreases. Where there are more than two parallel critical paths, the correction may take more than one cycle to fully correct.

Example

Consider compressing the duration of a project involving the construction of a single-span bridge. The relevant network is given in Figure 10.5, and the relevant cost–duration data and activity explanation are given in Table 10.3. In addition, an upper cost limit is assumed.

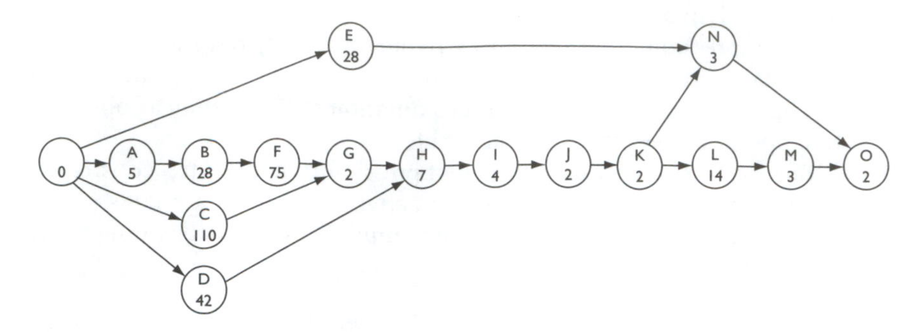

Figure 10.5 Single-span bridge construction network (Carmichael, 1989).

Table 10.3 Activity data for example. (Activity names have been abbreviated because of space limitations – work items are implied)

Activity Identification	*Activity Description*	*Normal*		*Crash*		*Cost/ Day*
		Dur'n (days)	*Cost*	*Dur'n* (days)	*Cost*	
A	Move-in	5	40.0	5	40.0	–
B	Coffer dam, divert river flow	28	60.0	24	65.0	1.25
C	O/d* bearings	110	20.0	90	50.0	1.5
D	O/d* precast girders	42	32.0	42	32.0	–
E	O/d* hand rails, accident barriers	28	6.0	28	6.0	–
F	Construct reinforced concrete abutments	75	100.0	70	130.0	6.0
G	Install bearings	2	4.0	2	4.0	–
H	Erect precast girders	7	10.0	6	12.0	2.0
I	Erect formwork for deck slab	4	3.0	3	5.5	2.5
J	Place deck slab reinforcement	2	4.0	2	4.0	–
K	Pour and finish deck slab concrete	2	2.0	2	2.0	–
L	Cure deck slab concrete	14	1.0	14	1.0	–
M	Lay bitumen surface	3	6.0	3	6.0	–
N	Install hand rails, accident barriers	3	3.0	3	3.0	–
O	Clean up	2	4.0	2	4.0	–

Note* O/d – Order and delivery.

The steps in the compression analysis are as follows:

Cycle 1
An initial network analysis assuming all normal durations gives the critical path through the activities O/d bearings (C) – Install bearings (G) – Erect girders (H) – Erect formwork (I) – Place reinforcement (J) – Pour slab – Cure concrete (K) – Lay bitumen (L) – Clean up (M). The project duration is 146 days.

Objective function

$$\min J = 1.5y_C + 2y_H + 2.5y_I$$

Constraints
Step constraints

$$y_C \geq 109, \quad y_H \geq 6, \quad y_I \geq 3$$

Activity duration ranges

$$90 \leq y_C \leq 110$$
$$6 \leq y_H \leq 7$$
$$3 \leq y_I \leq 4$$

Budget limit

$$295 + 1.5(YN_C - y_C) + 2.0(YN_H - y_H) + 2.5(YN_I - y_I) \leq 320$$

Solution

$$y_C = 109, \quad y_H = 6, \quad y_I = 3$$

Project duration = 143 days

Project direct cost = 301

Cycle 2
A reanalysis with updated activity durations for O/d bearings (C), Erect girders (H) and Erect formwork (I) shows that the critical path remains unchanged. Activities Erect girders (H) and Erect formwork (I) are at their crash values and can be removed from consideration.

Objective function

$$\min J = 1.5y_C$$

Constraints
Step constraints

$$y_C \geq 109$$

Activity duration ranges

$$90 \leq y_C \leq 110$$

Budget limit

$$295 + 1.5(YN_C - y_C) \leq 320$$

Solution

$$y_C = 108$$

Project duration = 142 days

Project direct cost = 302.5

Cycle 3
Parallel critical paths now exist with the path Move-in (A) – Coffer dam (B) – Construct abutments (F) also critical.

Objective function

$$\min J = 1.5y_C + 1.25y_B + 6y_F$$

Constraints
Step constraints

$$y_C \geq 107, \quad y_B \geq 25, \quad y_F \geq 71$$

Activity duration ranges

$$90 \leq y_C \leq 110$$

$$24 \leq y_B \leq 28$$

$$70 \leq y_F \leq 75$$

Budget limit

$$295 + 1.5(YN_C - y_C) + 1.25(YN_B - y_B) + 6(YN_F - y_F) \leq 320$$

where 320 is the assumed upper cost limit, and 295 is the all normal cost. (The all crash cost is 329.5.)

Solution

$$y_C = 107, y_B = 25, y_F = 71$$

Project duration = 141 days

Project direct cost = 311.25

Cycle 4

In the previous cycle, one of the parallel paths has been compressed by 2 days while the other has only been compressed by 1 day. That is, unnecessary expense has been incurred.

The parallel critical path has disappeared and the only remaining critical path is that with which the solution started.

This cycle rectifies the over compression of the previous cycle.

Solution

$$y_C = 106$$

Project duration = 140 days

Project direct cost = 312.75

Remaining cycles

The remaining cycles, with the budget limit removed, in addition to the early cycles reveal the progress shown in Table 10.4.

Note that the last cycle overshoots and adds cost without decreasing the project duration. The fourth, sixth, eighth and tenth cycles are corrections to over-compressing in the cycle preceding. Over-compressing is recognisable by there not being a full reduction in project duration commensurate with

Table 10.4 Sequence of project compressions

Cycle	*Activities compressed*	*Resulting project duration (days)*
		146
1	C, H, I	143
2	C	142
3	B, F, C	141
4	C	140
5	B, F, C	139
6	C	138
7	B, F, C	137
8	C	136
9	B, F, C	135
10	C	134
11	F, C	133
12	C	133

increased cost. Over-compressing then correction is a characteristic of the approach.

Example – Nonlinear cost functions

A generalisation of the problem solved in Deckro *et al.* (1992) is considered here.

The data remain the same as that given earlier but with the linear cost functions replaced by the following convex cost functions.

Activity	*Convex cost function*
A	$2209 - 23.9y + 1.11y^2$
B	$3370 - 35.0y + 2.00y^2$
C	$9385 - 16.9y + 0.57y^2$
D	–
E	$2430 - 45.0y + 0.75y^2$
F	$8564 - 32.0y + 3.00y^2$
G	$10030 - 46.6y + 0.94y^2$
H	$444 - 7.40y + 0.40y^2$
I	$7813 - 22.6y + 1.12y^2$
J	$9545 - 38.0y + 1.75y^2$
K	$9579 - 16.7y + 0.67y^2$

The linearised versions of these cost functions are obtained from a Taylor's series expansion. For example for activity A, this becomes

$$(2209 - 23.9y_A + 1.11y_A^2)\,|y^0 + (y_A - y_A^0)\left[-23.9 + 1.11y_A^0\right]$$

or

$$2209 + y_A\left[-23.9 + 1.11y_A^0\right]$$

For all activities, these are summarised as:

Activity	*Linear approximation to cost function*
A	$2209 + y_A(-23.9 + 1.11y_A^0)$
B	$3370 + y_B(-35.0 + 2.00y_B^0)$
C	$9385 + y_C(-16.9 + 0.57y_C^0)$
D	–
E	$2430 + y_E(-45.0 + 0.75y_E^0)$
F	$8564 + y_F(-32.0 + 3.00y_F^0)$
G	$10030 + y_G(-46.6 + 0.94y_G^0)$
H	$444 + y_H(-7.40 + 0.40y_H^0)$
I	$7813 + y_I(-22.6 + 1.12y_I^0)$
J	$9545 + y_J(-38.0 + 1.75y_J^0)$
K	$9579 + y_K(-16.7 + 0.67y_K^0)$

The solution develops as for the earlier flow diagram, using the linearised cost functions rather than the original nonlinear cost functions.

References and Bibliography

Butcher, W. S. (1967), Dynamic Programming for Project Cost-Time Curves, *Journal of the Construction Division, ASCE*, Vol. 93, No. CO1, March, pp. 59–73.

Carmichael, D. G. (1989), *Construction Engineering Networks*, Ellis Horwood Ltd (Wiley), Chichester.

Carmichael, D. G. (1996), An Iterative Optimisation Approach to Project Compression, in *Project Management Directions*, ed. D. G. Carmichael, B. Hovey, H. Rijsdijk and D. Baccarini, pp. 149–158, A Special Publication of the Australian Institute of Project Management, AIPM, Sydney.

Deckro, R. F., Hebert, J. E. and Verdini, W. A. (1992), Project Scheduling with Work Packages, *Omega*, Vol. 20, No. 2, pp. 169–182.

Fulkerson, D. R. (1961), A Network Flow Computation for Project Cost Curve, *Management Science*, Vol. 7, No. 2, pp. 167– 178.

Goya, S. K. (1975), A Note on 'A Simple CPM Time-Cost Trade-Off Algorithm', *Management Science*, Vol. 21, No. 6, February.

Kelly, J. E. (1961), Critical-Path Planning and Scheduling; Mathematical Basis, *Operations Research*, Vol. 9, pp. 296–320.

Meyer, W. L. and Shaffer, L. R. (1965), Extending CPM to Multiform Project Time-Cost Curves, *Journal of the Construction Division, ASCE*, Vol. 91, No. CO1, May, pp. 45–67.

Panagiotakopoulos, D. (1977), Cost-Time Model for Large CPM Project Networks, *Journal of the Construction Division, ASCE*, Vol. 103, No. CO2, June, pp. 201–211.

Perera, S. (1980), Linear Programming Solution to Network Compression, *Journal of the Construction Division, ASCE*, Vol. 106, No. CO3, September, pp. 315–326. Discussion, June 1981, pp. 409–410.

Perera. S. (1982), Compression of Overlapping Precedence Networks, *Journal of the Construction Division, ASCE*, CO1, March, pp. 1–12.

Prager, W. (1963), A Structural Method of Computing Project Cost Curves, *Management Science*, Vol. 9, pp. 394–404.

Robinson, D. R. (1975), A Dynamic Programming Solution to Cost-Time Trade-Off for CPM, *Management Science*, Vol. 22, No. 2, October.

Siemens, N. (1971), A Simple CPM Time-Cost Trade Off Algorithm, *Management Science*, Vol. 17, No. 6, February.

Index